Advances in Intelligent Systems and Computing

Volume 406

Series editor

Janusz Kacprzyk, Polish Academy of Sciences, Warsaw, Poland
e-mail: kacprzyk@ibspan.waw.pl

About this Series

The series "Advances in Intelligent Systems and Computing" contains publications on theory, applications, and design methods of Intelligent Systems and Intelligent Computing. Virtually all disciplines such as engineering, natural sciences, computer and information science, ICT, economics, business, e-commerce, environment, healthcare, life science are covered. The list of topics spans all the areas of modern intelligent systems and computing.

The publications within "Advances in Intelligent Systems and Computing" are primarily textbooks and proceedings of important conferences, symposia and congresses. They cover significant recent developments in the field, both of a foundational and applicable character. An important characteristic feature of the series is the short publication time and world-wide distribution. This permits a rapid and broad dissemination of research results.

Advisory Board

Chairman

Nikhil R. Pal, Indian Statistical Institute, Kolkata, India
e-mail: nikhil@isical.ac.in

Members

Rafael Bello, Universidad Central "Marta Abreu" de Las Villas, Santa Clara, Cuba
e-mail: rbellop@uclv.edu.cu

Emilio S. Corchado, University of Salamanca, Salamanca, Spain
e-mail: escorchado@usal.es

Hani Hagras, University of Essex, Colchester, UK
e-mail: hani@essex.ac.uk

László T. Kóczy, Széchenyi István University, Győr, Hungary
e-mail: koczy@sze.hu

Vladik Kreinovich, University of Texas at El Paso, El Paso, USA
e-mail: vladik@utep.edu

Chin-Teng Lin, National Chiao Tung University, Hsinchu, Taiwan
e-mail: ctlin@mail.nctu.edu.tw

Jie Lu, University of Technology, Sydney, Australia
e-mail: Jie.Lu@uts.edu.au

Patricia Melin, Tijuana Institute of Technology, Tijuana, Mexico
e-mail: epmelin@hafsamx.org

Nadia Nedjah, State University of Rio de Janeiro, Rio de Janeiro, Brazil
e-mail: nadia@eng.uerj.br

Ngoc Thanh Nguyen, Wroclaw University of Technology, Wroclaw, Poland
e-mail: Ngoc-Thanh.Nguyen@pwr.edu.pl

Jun Wang, The Chinese University of Hong Kong, Shatin, Hong Kong
e-mail: jwang@mae.cuhk.edu.hk

More information about this series at http://www.springer.com/series/11156

Alejandro Peña-Ayala
Editor

Mobile, Ubiquitous, and Pervasive Learning

Fundaments, Applications, and Trends

 Springer

Editor
Alejandro Peña-Ayala
WOLNM: Artificial Intelligence on
  Education Lab
Mexico City
Mexico

and

ESIME Zacatenco—Instituto Politécnico
  Nacional (IPN)
Mexico City
Mexico

ISSN 2194-5357   ISSN 2194-5365   (electronic)
Advances in Intelligent Systems and Computing
ISBN 978-3-319-26516-2   ISBN 978-3-319-26518-6   (eBook)
DOI 10.1007/978-3-319-26518-6

Library of Congress Control Number: 2015955361

Springer Cham Heidelberg New York Dordrecht London

Printed on acid-free paper

Springer International Publishing AG Switzerland is part of Springer Science+Business Media
(www.springer.com)

Preface

Mobile computing, ubiquitous computing, and pervasive computing are now synonymous for many, and represent a new model of computing in which computation is everywhere and computer functions are integrated into everything. Everyday objects will be placed for sensing, input, processing along with user output. In practice, this means the mobile, ubiquitous, and pervasive availability of many computing devices that are carried or embedded in homes, cars, offices, in walls and tables, or worn as part of smart clothing!

This book offers a glance into recent application of such computing paradigms in the field of education as a way for providing mobile, ubiquitous, and pervasive teaching–learning facilities for everybody, everywhere, and in any way. Thus, this book embraces several chapters that make up a sample of the work currently achieved in countries from four continents, which illustrates a sample of state of the art for the mobile, ubiquitous, and pervasive learning (MUP-Learning) arena. According to the nature of the contributions accepted for this volume, four kinds of topics are identified as follows:

- *Study* reports a research topic and the way it is tackled by the approach, where the MUP-Learning topic could be the target of research or just an instrument or scenery where the work is carried out.
- *Conceptual* describes a specific viewpoint that pursues to guide the design and development of an MUP-Learning approach by means of a model, method, or framework.
- *Review* provides a profile of a sample of works as well as some statistical analyses that shape a conceptual view of the labor recently carried out in the MUP-Learning field and its tendencies.
- *Approach* focuses on the application of a specific concept (e.g., technology, paradigm, tool, method…) to a learning setting where a complete research and development cycle is achieved since the research setting up to the experimental results.

This volume is the outcome of the research recently fulfilled by authors, who are willing to promote their models, methodologies, results, and findings to the community of practitioners, pedagogues, psychologists, computer scientists, academics, and students interested in the valuable topic of MUP-Learning!

As a result of the accomplishment of the cycle that embraces chapter submission, evaluation, decision, notification, and tuning according to the Springer quality principles, eight works were approved, edited as chapters, and organized according to the following sequence:

1. Chapter "The Effect of Question Styles and Methods in Quizzes Using Mobile Devices" is a study that evaluates the impact of question styles and methods. Thus, a battery of studies is reported to analyze differences between both smartphones and tablets, diverse question styles, series of questions and answers. One result reveals higher percentage of correct answers is achieved by tablet users than smartphones.

2. Chapter "A Generalized Approach for Context-Aware Adaptation in Mobile E-Learning Settings" is a conceptual work where a framework to build adaptive mobile learning applications is described. It claims the need to identify key contextual information to enable the design of a broad diversity of educational systems.

3. Chapter "A Revision of the Literature Concerned with Mobile, Ubiquitous, and Pervasive Learning: A Survey" offers a review of recent works carried out in the MUP-Learning field in order to shape a state of the art composed of empirical, conceptual, and domain-oriented approaches, as well as identify trends and challenges to tackle.

4. Chapter "Using Augmented Reality to Support Children's Situational Interest and Science Learning During Context-Sensitive Informal Mobile Learning" corresponds to an approach aimed at using augmented reality to stimulate children situational interest in science during context-sensitive informal learning. A case study is explored where children learn about trees by means of using iPads and augmented reality stimulates situational interest.

5. Chapter "Cooperative Face-to-Face Learning with Connected Mobile Devices: The Future of Classroom Learning?" is an approach to enhance cooperative face-to-face learning by playing a learning game for iPhone/iPad devices. The results reveal the usefulness of the application and how it motivates children to learn math.

6. Chapter "Prospective Teachers—Are They Already Mobile?" is a study to explore how teachers are willing to use mobile phones and laptops in the classroom. Thus, a cross-sectional survey is performed among 650 prospective Turkish teachers. One conclusion claims the need to motivate teachers to become aware of the mobile learning potential to enhance class.

7. Chapter "Flexible and Contextualized Cloud Applications for Mobile Learning Scenarios" represents a conceptual work aimed at providing a series of guidelines to encourage the development of mobile sceneries in cloud computing

environments. So, a framework for teachers interested in building mobile learning systems, and a service to deploy personalized environments are stated.

8. Chapter "Toward an Adaptive and Adaptable Architecture to Support Ubiquitous Learning Activities" offers a review of the notion of adaptive and adaptable architecture suitable to develop ubiquitous learning systems. As result, a model of a domain-specific architecture is designed as a baseline for building applications.

I express my gratitude to authors, reviewers of the Springer editorial team, and the editors Dr. Thomas Ditzinger and Prof. Janusz Kacprzyk for their valuable collaboration to fulfill this work.

I also acknowledge the support given by the Consejo Nacional de Ciencia y Tecnología (CONACYT) and the Instituto Politécnico Nacional (IPN) of Mexico through the grants: CONACYT-SNI-36453, IPN-SIP-20150910, IPN-SIP-EDI-848-14, IPN-COFAA-SIBE-ID: 9020/2015-2016, CONACYT 264215.

Last but not least, I acknowledge the strength given by my Father, Brother Jesus, and Helper, as part of the research projects of World Outreach Light to the Nations Ministries (WOLNM).

September 2015   Alejandro Peña-Ayala

Contents

The Effect of Question Styles and Methods in Quizzes Using Mobile Devices 1
Takeshi Kitazawa, Koki Sato and Kanji Akahori

A Generalized Approach for Context-Aware Adaptation in Mobile E-Learning Settings 23
Tobias Moebert, Raphael Zender and Ulrike Lucke

A Revision of the Literature Concerned with Mobile, Ubiquitous, and Pervasive Learning: A Survey 55
Alejandro Peña-Ayala and Leonor Cárdenas

Using Augmented Reality to Support Children's Situational Interest and Science Learning During Context-Sensitive Informal Mobile Learning 101
Heather Toomey Zimmerman, Susan M. Land and Yong Ju Jung

Cooperative Face-to-Face Learning with Connected Mobile Devices: The Future of Classroom Learning? 121
Martin Ebner, Sandra Schön, Hanan Khalil and Barbara Zuliani

Prospective Teachers—Are They Already Mobile? 139
Süleyman Nihat Şad, Özlem Göktaş and Martin Ebner

Flexible and Contextualized Cloud Applications for Mobile Learning Scenarios 167
Alisa Sotsenko, Janosch Zbick, Marc Jansen and Marcelo Milrad

Toward an Adaptive and Adaptable Architecture to Support Ubiquitous Learning Activities 193
Janosch Zbick, Bahtijar Vogel, Daniel Spikol, Marc Jansen and Marcelo Milrad

Author Index 223

Contributors

Kanji Akahori JAPET: Japan Association for Promotion of Educational Technology, Minato-ku, Tokyo, Japan; CRET: Center for Research on Educational Testing, Shinjuku-ku, Tokyo, Japan

Leonor Cárdenas ESIME Zacatenco, Instituto Politécnico Nacional, Mexico, DF, Mexico

Martin Ebner Educational Technology, Graz University of Technology, Graz, Austria

Özlem Göktaş Ministry of National Education, Sumer Secondary School, Yesilyurt, Malatya, Turkey

Marc Jansen Department of Media Technology, Linnaeus University, Växjö, Sweden; University of Applied Sciences Ruhr West, Mülheim an der Ruhr, Germany

Yong Ju Jung Learning, Design and Technology Program, Penn State University, PA, USA

Hanan Khalil Manosoura University, Mansoura, Egypt

Takeshi Kitazawa Department of Technology and Information Science, Tokyo Gakugei University, Koganei-shi, Tokyo, Japan; CRET: Center for Research on Educational Testing, Shinjuku-ku, Tokyo, Japan

Susan M. Land Learning, Design and Technology Program, Penn State University, PA, USA

Ulrike Lucke Universität Potsdam, Institut für Informatik, Potsdam, Germany

Marcelo Milrad Department of Media Technology, Linnaeus University, Växjö, Sweden

Tobias Moebert Universität Potsdam, Institut für Informatik, Potsdam, Germany

Alejandro Peña-Ayala WOLNM: Artificial Intelligence on Education Lab, Mexico City, Mexico; ESIME Zacatenco—Instituto Politécnico Nacional (IPN), Mexico City, Mexico

Süleyman Nihat Şad İnönü University, Curriculum and Instruction, Malatya, Turkey

Koki Sato International Language Center, International Education & Exchange, Nagoya University, Chikusa-ku, Nagoya, Japan

Sandra Schön Salzburg Research, Salzburg, Austria

Alisa Sotsenko LNU: Linnaeus University, Växjö, Sweden

Daniel Spikol Department of Media Technology, Malmö University, Malmö, Sweden

Bahtijar Vogel Department of Media Technology, Malmö University, Malmö, Sweden

Janosch Zbick Department of Media Technology, Linnaeus University, Växjö, Sweden

Raphael Zender Universität Potsdam, Institut für Informatik, Potsdam, Germany

Heather Toomey Zimmerman Learning, Design and Technology Program, Penn State University, PA, USA

Barbara Zuliani Elementary School Breitenlee, Vienna, Austria

The Effect of Question Styles and Methods in Quizzes Using Mobile Devices

Takeshi Kitazawa, Koki Sato and Kanji Akahori

Abstract In this chapter, quizzes were administered to university students using smartphones and tablets. The impact of question styles and methods on motivation, and the percentage of correct answers were then investigated, while considering the test approach–avoidance tendencies. In Study 1, 20 multiple-choice questions were set, and differences between smartphones and tablets were analyzed in terms of the percentage of correct answers and the optimum number of questions. In Study 2, different question styles, namely multiple-choice, fill-in-the-blank, and a combination of both multiple choice and fill-in-the-blank, using tablets, were analyzed. In Study 3, for the question method for quizzes using smartphones, all questions were displayed and a series of questions and answers were analyzed for comparison. In Study 4, for the question method for quizzes, which was considered in the context of using smartphones, all questions were displayed and a series of questions and answers were analyzed.

Keywords Smartphone · Tablet computer · Question style · Quiz · Motivation · Test approach–avoidance tendency · Self-efficacy

---

T. Kitazawa (✉)
Department of Technology and Information Science, Tokyo Gakugei University, 4-1-1 Nukuikita-machi, Koganei-shi, Tokyo 184-8501, Japan
e-mail: ktakeshi@u-gakugei.ac.jp

K. Sato
International Language Center, International Education & Exchange, Nagoya University, Furo-cho, Chikusa-ku, Nagoya 464-8601, Japan
e-mail: sato@iee.nagoya-u.ac.jp

K. Akahori
JAPET: Japan Association for Promotion of Educational Technology, 1-9-13 Akasaka, Minato-ku, Tokyo 107-0052, Japan
e-mail: akahori@japet.or.jp

T. Kitazawa · K. Akahori
CRET: Center for Research on Educational Testing, Shinjuku Mitsui Bldg., 13F, 2-1-1, Nishi Shinjuku, Shinjuku-ku, Tokyo 163-0413, Japan

© Springer International Publishing Switzerland 2016
A. Peña-Ayala (ed.), *Mobile, Ubiquitous, and Pervasive Learning*, Advances in Intelligent Systems and Computing 406, DOI 10.1007/978-3-319-26518-6_1

Abbreviation

IT   Information technologies

1   Introduction

As a result of faculty development being made compulsory, Japanese universities have been asked to improve their classes with the aim of promoting student learning outside of class (i.e., in after-class hours) as well as improving student academic achievement and knowledge retention [1]. Blended learning, which combines the e-learning system with computers and face-to-face approaches, has focused on one of the learning environments to increase university students' learning activities [2].

However, a seamless learning environment is needed to develop students' learning outside of school hours [3, 4]. Blended learning that combines the "m-learning" system with mobile devices and face-to-face approaches is more suitable to develop seamless learning [5]. It is anticipated that constructing an environment in which students can access an e-learning system using mobile devices will encourage after-class learning [6], as it will allow students to engage in learning anywhere and at any time [7–11]. It is assumed that the students are able to use mobile devices more easily than desktop computers because almost all Japanese university students have their own smartphones. It is important for us to focus on smartphones and tablet computers as mobile devices for ubiquitous learning, so that students are able to motivate themselves to study in after-class hours [12, 13]. Therefore, we have focused on blended learning that combines the m-learning system with smartphone and tablet computers and face-to-face approaches as an optimum learning environment to increase university students' learning activities [14, 15].

One of the purposes of faculty development is to increase students' knowledge. Administering tests effectively can be considered a method of knowledge retention [16]. Based on this finding, one can expect that asking students to take tests in a mobile environment outside of class hours will produce a knowledge-retention effect and promote after-class learning [17]. However, research on mobile learning has not focused on smartphones and tablet computers, but on mobile phones [7–11]. Research on m-learning that focuses on using smartphones and tablet computers is therefore needed [13]. For this reason, we have researched how to use these devices for quizzes.

Since learners' perception of tests, or their perception of the objective of conducting tests and their roles, affects their learning behaviors, it is important to administer tests that take into account learners' approach–avoidance tendencies [18]. Additionally, the students' motivation towards the tests influences the results of the quizzes in a positive and negative way. Therefore, we first investigated students' test approach–avoidance tendencies, before they took the quizzes on their smartphone and tablet computers.

The purpose of this chapter is to describe the participants of analysis, who were students enrolled in an Introduction to Information Science university course, with an experimental environment being developed to send quizzes to them during after-class hours on the assumption that they would do the quiz independently. The reason why we targeted the course was its focus on information technology (IT), a subject that had become compulsory for high schools in 2003, with the number of subjects along with its teaching content having been revised in the academic year of 2013. Therefore, it is necessary to improve information education classes that are aimed at new students enrolled in information technologies (IT) courses.

We will describe three studies to clarify the relationship among the mobile devices (smartphones or tablet computers), the question styles and methods in quizzes, and the results of answers from quizzes in terms of students' test approach–avoidance tendencies based on the authors' prior research [19]. In addition, we will report one new study that was researched to further elaborate on the tasks in our previous research.

Finally, it is necessary to acknowledge that the research described in this chapter takes into account the work carried out by authors that has been partially published in the references [14, 15, 19, 20].

2 Study 1

In Study 1, twenty multiple-choice questions were set, and the differences between smartphones and tablet computers were analyzed in terms of the percentage of correct answers and the optimum number of questions (Fig. 1).

2.1 Participants

The participants comprised 22 Japanese university students. The research was conducted on September 8, 2012.

2.2 Administering Quizzes to Tablets and Smartphones

After the lecture, 11 students took a quiz comprising 20 multiple-choice questions on a tablet terminal, while 11 students took the same test using a smartphone (Fig. 2). The quizzes were distributed during after-class hours. The tablet terminal used in this study was an iPad (1st generation), 16 GB, Wi-Fi model. The smartphone used was an iPod touch (4th generation), 64 GB, Wi-Fi model.

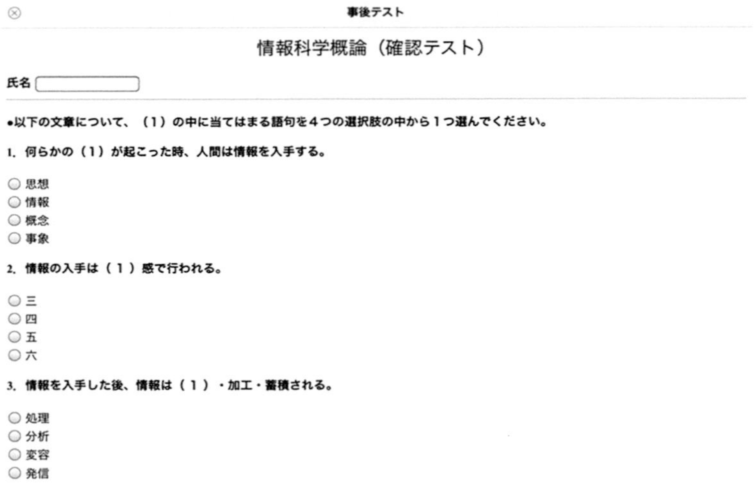

⊗   事後テスト

情報科学概論（確認テスト）

氏名 [　　　　　　　　]

●以下の文章について、（1）の中に当てはまる語句を4つの選択肢の中から1つ選んでください。

1. 何らかの（1）が起こった時、人間は情報を入手する。

○ 思想
○ 情報
○ 概念
○ 事象

2. 情報の入手は（1）感で行われる。

○ 三
○ 四
○ 五
○ 六

3. 情報を入手した後、情報は（1）・加工・蓄積される。

○ 処理
○ 分析
○ 変容
○ 発信

Fig. 1 Sample of multiple-choice questions [19]

Fig. 2 Aspect of Study 1 (*left* tablet computer user, *right* example of smartphone use)

2.3 Study Design

Table 1 indicates the design of Study 1. First, we performed a pretest on test approach–avoidance tendencies using a 7-item scale [18]. Second, the students attended a lecture on the introductory section of the course "Introduction to Information Science," for example, "What is information?" and "What is the definition of information?" The lecture lasted approximately 15 min. They could scribble notes on a worksheet. Third, a few minutes after the lecture, 11 students

Table 1 Study design of Study 1	1. The university students had a pretest on test approach–avoidance tendencies
	2. They attended a lecture on the introductory section of the course "Introduction to Information Science"
	3. 11 students took a quiz comprising 20 multiple-choice questions using a tablet terminal, while 11 students took the same test using a smartphone. Authors analyzed the percentage of correct answers
	4. They had a posttest on the recognition of the quiz, motivation towards the quiz, self-efficacy, and the device they used. The authors analyzed the results of the questionnaire by statistical methods

took a quiz comprising 20 multiple-choice questions using a tablet terminal, while 11 students took the same test using a smartphone. They were not able to check the worksheet while they took the quizzes. Finally, they had a posttest, using a 5-item scale, about their recognition of the quiz, motivation towards the quiz, self-efficacy, and the device they used.

We analyzed the percentage of correct answers about the quizzes in terms of test approach–avoidance tendency. To analyze the tendency, the test approach–avoidance tendency scale (a 7-item scale with 10 questions) developed by Suzuki [18] was used as a pre-questionnaire. A post-questionnaire survey (a 5-item scale with 15 questions) was administered to assess the students' "burden of taking quizzes," "willingness to have quizzes," and "self-efficacy," with the aim of understanding their motivation regarding quizzes. The data were analyzed by an overall trend analysis, comparing them with the median values.

2.4 Results

We categorized test approach–avoidance tendencies into four groups, namely "high–high," "high–low," "low–high," and "low–low." The results of the test approach–avoidance tendencies' questionnaire were as follows: smartphone group had a high–high of five persons, high–low of one person, low–high of three persons, and low–low of two persons. The tablet computer group had a high–high of two persons, high–low of three persons, low–high of three persons, and low–low of two persons.

We analyzed the data by the Kruskal–Wallis test. The results of the percentage of correct answers were as follows: the smartphone group was "high–high (93.7 %, SD = 0.04)," "high–low (94.7 %, SD = 0.00)," "low–high (8.9 %, SD = 0.07)," and "low–low (89.5 %, SD = 0.05);" the tablet computer group was "high–high (81.6 %, SD = 0.18)," "high–low (94.7 %, SD = 0.04)," "low–high (91.2 %,

SD = 0.05)," and "low–low (97.4 %, SD = 0.03)." There were no significant differences between smartphone and tablet computers with regard to the percentage of correct answers (χ^2 (7) = 6.80, *n.s.*).

From the result of the posttest, there were no significant differences between smartphones and tablet computers with regard to the recognition of the quiz, motivation towards the quiz, and self-efficacy. But an item of recognition about the device they used which was analyzed using the Mann–Whitney U test showed significant differences, as follows: recognition about the number of the quizzes (the average of the smartphone group was 3.00 (SD = 0.00), while the average of the tablet computer group was 3.30 (SD = 0.46), $p < 0.10$).

From the results of the average of the students' answers about "the number of the questions that I am willing to do in a quiz," with the smartphone group at 13.6 and the tablet computers' group at 18.3, we found that the suitable number of questions in a quiz was 15.

3   Study 2

In Study 2, different question styles, namely multiple choices, fill-in-the-blank, and a combination of both multiple choice and fill-in-the-blank, were set, with the students taking the quizzes on tablet computers (Fig. 3). The differences in question styles in terms of student motivation and the percentage of correct answers were thus analyzed.

3.1   Participants

The participants comprised 60 Japanese university students. The research was conducted on October 14, 2012. The procedures of the class and taking the quizzes were the same as Study 1.

3.2   Administering Quizzes by Tablet Computers

After the lecture, 60 students took quizzes comprising 10 multiple-choice questions, 10 fill-in-the-blank questions, and 10 combinations of both multiple choice and fill-in-the-blank questions, using a tablet terminal. The quizzes were distributed during after-class hours. The tablet terminal used in this study was an iPad (3rd generation), 16 GB, Wi-Fi model.

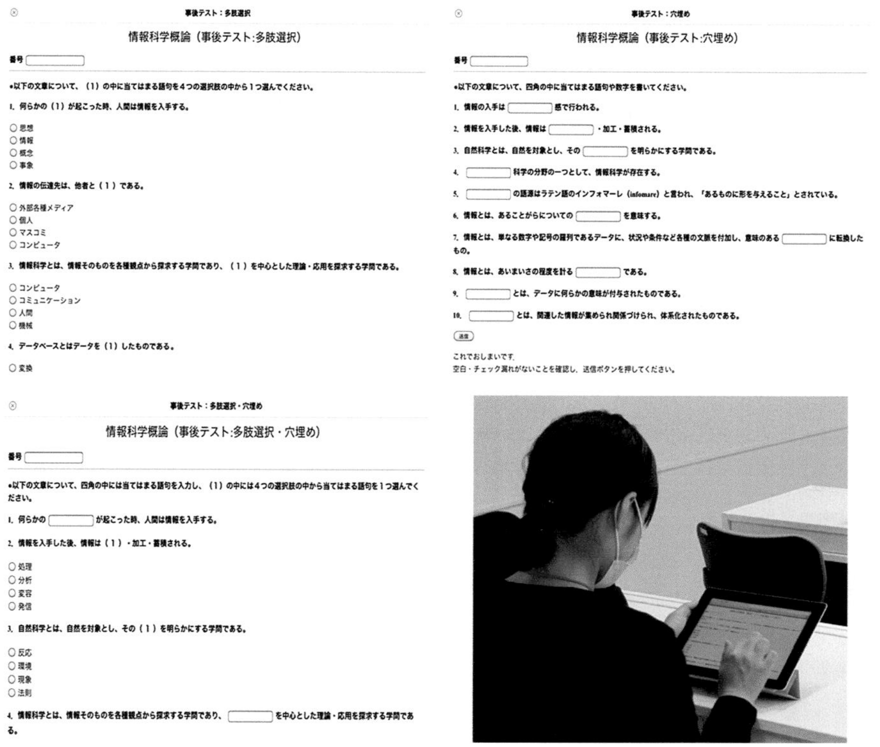

Fig. 3 Sample of quizzes (*upper left* multiple-choice questions; *upper right* fill-in-the-blank questions; *left below* combinations of both multiple choice and fill-in-the-blank questions) [19]. *Right below* is an aspect of Study 2

3.3 Study Design

Table 2 shows the design of Study 2. First, we made a pretest with a 7-item scale about the test approach–avoidance tendencies [18]. Second, students attended a lecture on the introductory section of the course "Introduction to Information Science," for example, "What is information?" and "What is the definition of information" in a similar way to Study 1. Third, a few minutes after the lecture, 60 students took three quizzes comprising 10 multiple-choice questions, 10 fill-in-the-blank questions, and 10 combinations of both multiple choice and fill-in-the-blank questions.

We divided them into three groups and considered the order effect. The 20 students of Group 1 took the quizzes in the order corresponding to multiple-choice questions, fill-in-the-blank questions, and combinations of both multiple choice and fill-in-the-blank questions. The 20 students of Group 2 took the quizzes in the order

Table 2 Study design of Study 2	1. The university students had a pretest on test approach–avoidance tendencies
	2. They attended a lecture on the introductory section of the course "Introduction to Information Science"
	3. 60 students took three quizzes comprising 10 multiple-choice questions, 10 fill-in-the-blank questions, and 10 combinations of both multiple choice and fill-in-the-blank questions. Authors analyzed the percentage of correct answers
	4. They had a posttest on their recognition of the quiz, motivation towards the quiz, self-efficacy, and the device they used. Authors analyzed the results of the questionnaire by statistical methods

corresponding to fill-in-the-blank questions, combinations of both multiple choice and fill-in-the-blank questions, and multiple-choice questions.

As for the 20 students of Group 3, they took the quizzes in the order corresponding to combinations of both multiple choice and fill-in-the-blank questions, multiple-choice questions, and fill-in-the-blank questions.

Finally, they had a posttest, using a 5-item scale, on their recognition of the quiz, motivation towards the quiz, self-efficacy, and the device they used.

3.4 Results

The results of the test approach–avoidance tendencies' questionnaire were as follows: "high–high" had 17 persons, "high–low" had 16 persons, "low–high" had 16 persons, "low–low" had 9 persons, and the missing value had 2 persons. The tablet computer group had two high–high persons, three high–low persons, three low–high persons, and two low–low persons.

We compared the test approach–avoidance tendencies (four categories) and question styles (three categories) by means of two-way ANOVA. We found that there was an interaction significantly different between the test approach–avoidance tendencies and question styles ($F(6108) = 2.47$, $p < 0.05$). In addition, the main effect of the question styles was recognized to be significantly different ($F(2108) = 52.0$, $p < 0.01$). With regard to the average of the correct answers of multiple-choice questions, "low–low" was 80.0 (SD = 16.5) and "high–low" was 84.4 (SD = 14.6). In contrast, "low–low" was 71.1 (SD = 17.6) and "high–low" was 48.8 (SD = 20.9) for the fill-in-the-blank questions. These results show an interaction between test approach–avoidance tendencies and question styles.

Next, we focus on the results of Tukey multiple comparisons on the main effect. There was a significant difference between the multiple-choice questions (high–high was $M = 76.5$, SD = 13.7, high–low was $M = 84.4$, SD = 14.6, low–high was $M = 76.9$, SD = 21.5, and low–low was $M = 80.0$, SD = 16.6)" and the "fill-in-the-blank questions (high–high was $M = 52.9$, SD = 17.2, high–low was

$M = 48.8$, SD = 20.9, low–high was $M = 55.6$, SD = 20.0, low–low was $M = 71.1$, SD = 17.6) ($p < 0.01$). Additionally, there was a significant difference between the results of the "fill-in-the-blank questions" and "combinations of both multiple choice and fill-in-the-blank questions (high–high was $M = 71.8$, SD = 13.8, high–low was $M = 74.4$, SD = 17.9, low–high was $M = 73.1$, SD = 23.0, low–low was $M = 87.8$, SD = 13.0) ($p < 0.01$)."

Regardless of the student test approach–avoidance tendencies, the percentage of correct answers for the fill-in-the-blank questions was lower than the multiple-choice questions and a combination of both multiple choice and fill-in-the-blank. A combination of both multiple choice and fill-in-the-blank was the best quiz style of the three, so that the recognition on self-efficacy of the style was higher than that of the multiple choice questions, and the percentage of correct answers about the style was higher than the fill-in-the-blank questions.

From the results of the posttest by two-way ANOVA, we found that the items of "burden about doing quizzes" had a main effect ($F(2, 108) = 93.0$, $p < 0.01$). From the results of simple main effects and Tukey multiple comparisons, we found that the students recognized the greater burden of fill-in-the-blank questions than multiple-choice questions, and combinations of both multiple choice and fill-in-the-blank questions. The items about "motivation regarding quiz styles" had not only an interaction of ($F(24, 424) = 1.90$, $p < 0.01$) but also a main effect of ($F(8, 424) = 18.6$, $p < 0.01$).

The results indicate that students recognized being more motivated by combinations of both multiple choice and fill-in-the-blank questions than fill-in-the-blank questions alone, despite their test approach–avoidance tendencies.

From the results of the percentage of correct answers and recognition of the students, we found that the quiz style of combinations of both multiple choice and fill-in-the-blank questions was the best of the three styles.

4 Study 3

As for Study 3, the question method for quizzes using smartphones was taken into account, where students were divided into two groups, with either all questions being displayed or a series of questions and answers (Fig. 4). A comparative group analysis was conducted in terms of student motivation and the percentage of right answers.

4.1 Participants

The participants comprised 60 Japanese university students. They used their own smartphones (Fig. 5). The research was conducted on October 5, 2013. The procedures of the class and taking the quizzes were the same as Studies 1 and 2.

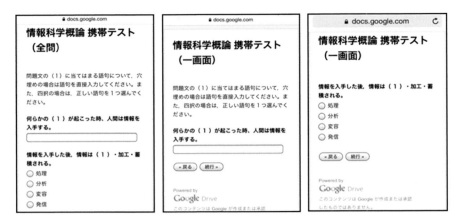

Fig. 4 Sample of quizzes (*left* all questions being displayed; *center* and *right* a series of questions and answers) [19]

Fig. 5 Aspect of Study 3

4.2 Administering Quizzes by Smartphone

After the lecture, 60 students took quizzes comprising 15 combinations of both multiple choice and fill-in-the-blank questions, using smartphones. We adopted such quizzes because of the findings of Studies 1 and 2. The quizzes were distributed during after-class hours.

4.3 Study Design

Table 3 indicates the design of Study 3. First, we performed the pretest with a 7-item scale about test approach–avoidance tendencies [18]. Second, the students attended a lecture on the introductory section of the course "Introduction to Information Science," for example, "What is information?" and "What is the definition of information" in a similar way to Studies 1 and 2. Third, a few minutes after the lecture, 60 students took quizzes comprising 15 combinations of both multiple choice and fill-in-the-blank questions. However, 30 students took the quizzes with all questions displayed, and 30 students took the quizzes with a series of questions and answers.

Finally, they had a posttest with a 5-item scale about their recognition of the quiz, motivation towards the quiz, self-efficacy, and the device they used.

4.4 Results

The results of the test approach–avoidance tendencies' questionnaire were as follows: the "all questions displayed" group had a high–high of 7 persons, high–low of 11 persons, low–high of nine persons, and low–low of 3 persons. The "series of questions and answers" group had a high–high of 8 persons, high–low of 7 persons, low–high of 5 persons, and low–low of 10 persons.

From the results of two-way ANOVA, we found that the methods of the quiz had a main effect of ($F(1, 52) = 9.46$, $p < 0.01$). From the results of a Tukey multiple comparison, the percentage of correct answers about the quiz with all questions displayed ($M = 76.0$, SD $= 16.5$) was higher than the quiz with a series of questions and answers ($M = 64.0$, SD $= 15.2$). Additionally, the percentage of correct answers about "low–low" had a main effect ($F(1, 52) = 6.34$, $p < 0.05$). From the results of a Tukey multiple comparison, the percentage of correct answers about the quiz with all questions displayed of "low–low ($M = 88.9$, SD $= 10.2$)"

Table 3 Study design of Study 3	1. The university students had a pretest on test approach–avoidance tendencies
	2. They attended a lecture on the introductory section of the course "Introduction to Information Science"
	3. 30 students took the quizzes with all questions displayed, and 30 students took the quizzes with a series of questions and answers. The authors analyzed the percentage of correct answers
	4. They had a posttest on their recognition of the quiz, motivation towards the quiz, self-efficacy, and the device they used. The authors analyzed the results of the questionnaire by statistical methods

was higher than that of the quiz of a series of questions and answers displayed one after the other ($M = 62.0$, SD $= 16.6$). We consider the reason for the percentage of correct answers on the quiz with all questions displayed to be higher because it was easier for the students to understand the context of the quizzes.

The time for taking the quiz showed no significant differences. The students who took the quiz with a series of questions and answers recognized that it was easier for them than the quiz with all questions displayed. Especially, the students whose test approach–avoidance tendencies were a high–low score strongly recognized that they were not willing to have the quiz with all questions displayed. On the other hand, the students who took the quiz with all questions displayed tended to recognize self-efficacy, such as understanding the quiz.

From the results of "burden of taking quizzes" of the posttest, Question 4, "It was a burden to take the quiz that had all questions displayed rather than a series of questions and answers" had main effects both with quiz method ($F(1, 52) = 9.90$, $p < 0.01$) and test approach–avoidance tendencies ($F(1, 52) = 3.23$, $p < 0.05$). According to the results of simple main effects and Tukey multiple comparison, there was a significant difference between the "all questions displayed" group ($M = 2.53$, SD $= 0.90$) and the "series of questions and answers" group ($M = 3.17$, SD $= 0.99$) ($p < 0.01$). This finding shows that the students who took the quiz that had the series of questions and answers tended to recognize the burden of taking quizzes with all questions displayed. Furthermore, when we focused on the test approach–avoidance tendencies of students, we found that "high–low" students who took the quizzes that had a series of questions and answers recognized it as higher ($M = 3.14$, SD $= 1.07$) than those who took the quizzes which had all questions displayed ($M = 2.18$, SD $= 0.87$) ($p < 0.05$). Therefore, the students who are motivated towards the test may find all questions displayed a burden, having experienced them as a series of questions and answers.

Next, we focus on the results of "self-efficacy." Question 19, "I understood what I learned in this class enough to take the quiz that has all questions displayed rather than a series of questions and answers." had a main effect of ($F(1, 52) = 5.68$, $p < 0.05$). The average of the "all questions displayed" group was $M = 2.73$, SD $= 0.83$. The average of the "series of questions and answers" group was $M = 2.27$, SD $= 0.87$. This result and the percentage of correct answers indicate that the quiz method of all questions being displayed is better than a series of questions and answers in terms of understanding. This finding may be related to understanding the context of quizzes, which is a necessary are of future research.

5   Study 4

From the results of Study 3, in Study 4 we focused on the context of the quizzes. Students were divided into two groups, with either all questions displayed or a series of questions and answers, for which the questions had been improved with context. A comparative group analysis was conducted in terms of student motivation and the percentage of correct answers.

5.1 Participants

The participants comprised 60 Japanese university students. The research was conducted on September 27, 2014. The procedures of the class and taking the quizzes were the same for Studies 1, 2, and 3.

5.2 Administering Quizzes by Smartphone

After the lecture, 60 students took the quizzes comprising 15 combinations of both multiple choice and fill-in-the-blank questions, using smartphones. The quizzes were distributed during after-class hours. The quizzes had been improved considering the context. The quizzes of Studies 1, 2, and 3 were one-to-one correspondences between a question and an answer. Although the questions and answers were the same as Studies 1, 2, and 3 (Fig. 6), the questions of the quiz of Study 4 were contextualized.

5.3 Study Design

Table 4 indicates the design of Study 4. First, we performed a pretest with a 7-item scale about test approach–avoidance tendencies [18]. Second, the students attended a lecture on the introductory section of the course "Introduction to Information Science," for example, "What is information?" and "What is the definition of information" in a similar way to Studies 1, 2, and 3. Third, a few minutes after the lecture, 60 students took quizzes comprising 15 combinations of both multiple choice and fill-in-the-blank questions in consideration of the context. 30 students took the quizzes with all questions displayed, and 30 students took the quizzes with a series of questions and answers (Fig. 7).

Finally, they had a posttest with a 5-item scale about their recognition of the quiz, motivation towards the quiz, self-efficacy, and the device they used.

5.4 Results

The results of test approach–avoidance tendencies' questionnaire were as follows: the "all questions displayed" group had a high–high of 3 persons, high–low of 16

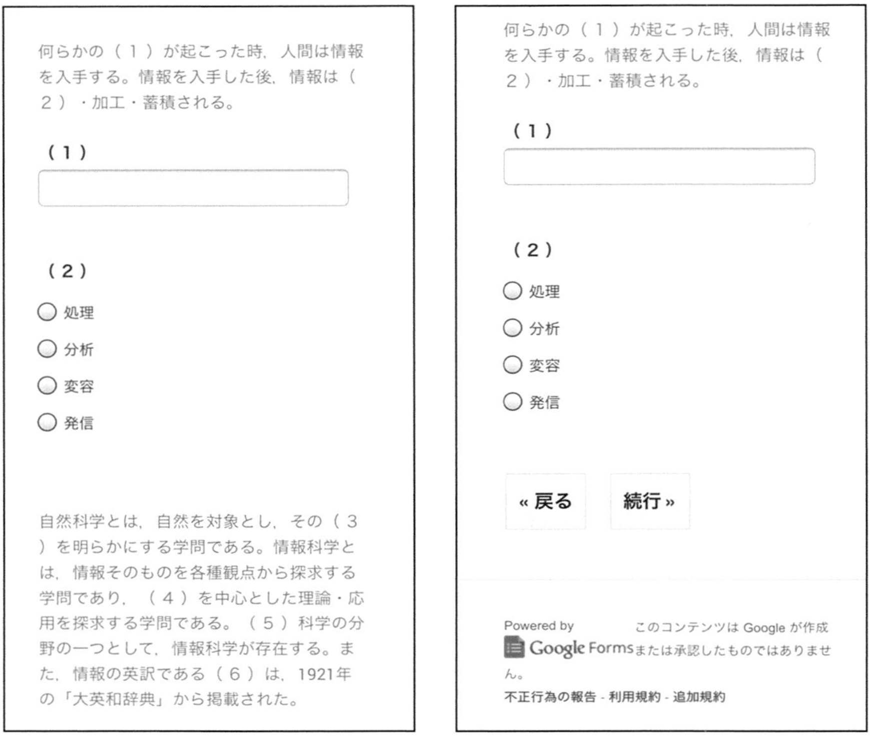

何らかの（ 1 ）が起こった時，人間は情報
を入手する。情報を入手した後，情報は（
2 ）・加工・蓄積される。

（ 1 ）

［   ］

（ 2 ）

○ 処理

○ 分析

○ 変容

○ 発信

自然科学とは，自然を対象とし，その（ 3
）を明らかにする学問である。情報科学と
は，情報そのものを各種観点から探求する
学問であり，（ 4 ）を中心とした理論・応
用を探求する学問である。（ 5 ）科学の分
野の一つとして，情報科学が存在する。ま
た，情報の英訳である（ 6 ）は，1921年
の「大英和辞典」から掲載された。

何らかの（ 1 ）が起こった時，人間は情報
を入手する。情報を入手した後，情報は（
2 ）・加工・蓄積される。

（ 1 ）

［   ］

（ 2 ）

○ 処理

○ 分析

○ 変容

○ 発信

《 戻る   続行 》

Powered by   このコンテンツは Google が作成
Google Forms または承認したものではありませ
ん。

不正行為の報告 - 利用規約 - 追加規約

Fig. 6 Sample of quizzes in consideration of context (*left* all questions displayed; *right* a series of questions and answers) [20]

Table 4 Study design of Study 4	1. The university students had a pretest on test approach–avoidance tendencies
	2. They attended a lecture on the introductory section of the course "Introduction to Information Science"
	3. 30 students took the quizzes with added context all questions displayed, and 30 students took the quizzes in consideration of the context with the series of questions and answers. The authors analyzed the percentage of correct answers
	4. They had a posttest on their recognition of the quiz, motivation towards the quiz, self-efficacy, and the device they used. The authors analyzed the results of the questionnaire by statistical methods

persons, low–high of 6 persons, low–low of 4 persons, and the missed value was 1 person. The "series of questions and answers" group had a high–high of nine persons, high–low of seven persons, low–high of eight persons, and low–low of six persons.

Fig. 7 Aspect of Study 4

From the results of two-way ANOVA, we found that the methods of quizzes had no main effect ($F(1, 51) = 1.69$, *n.s.*). But test approach–avoidance tendencies had a main effect ($F(3, 51) = 6.41$, $p < 0.01$). From the results of a Tukey multiple comparison, it was clear that "low–low ($M = 60.0$, SD $= 10.4$)" was lower than "high–low ($M = 72.2$, SD $= 8.2$) ($p < 0.05$) (Fig. 8)."

This result shows that the percentage of correct answers of the students whose test approach is high tend to become higher even if the test avoidance tendency is lower. In addition, there was a significant difference between "low–high ($M = 55.2$)"

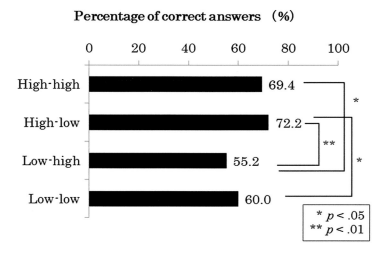

Fig. 8 Results of the quiz [20]

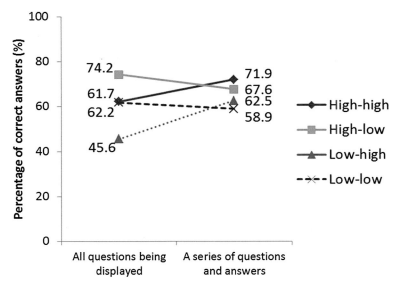

Fig. 9 Results of the quizzes compared by quiz methods [20]

and "high–high ($M = 69.4$, SD = 16.9) ($p < 0.05$)." From these results, we found that the percentage of the correct answers was affected by the test approach tendency.

We focused on results of the quizzes compared by quiz methods (Fig. 9). There was an interaction between quiz method and test approach–avoidance tendencies ($F(3, 51) = 3.30$, $p < 0.05$).

From the results of simple main effects, it can be seen that "low–high" had a significant difference by quiz method ($F(1, 51) = 7.58$, $p < 0.01$). From the results of a Tukey multiple comparison, it was clear that the "series of questions and answers" group ($M = 45.6$) was higher than the "all questions displayed" group ($M = 62.5$) ($p < 0.01$).

These findings show that "low–high" students who were negative towards the tests adapted to the "series of questions and answers" method rather than the "all questions displayed" method. In other words, "low–high" students seem to prefer taking a test one question at a time rather than taking many questions at the same time.

6   Future Trends and Conclusions

In this chapter, quizzes were given to students enrolled in a university information foundation course outside of the classroom, using mobile devices such as smart-phones and tablet computers. The impacts of question styles and methods on motivation and the percentage of correct answers were then investigated, while taking into account the test approach–avoidance tendencies of the university students.

This chapter suggests what kinds of question styles and quiz formats are better for them. The results of studies in this chapter indicate how to improve mobile learning approaches. The details of the findings are as follows:

- In Study 1, there was no significant difference between smartphone users and users of tablet computers in terms of the percentage of correct answers of the 20 multiple-choice questions.
- In Study 1, averaging the results of the students' answers on "the number of the questions that I am willing to do in the quiz," with the smartphone group at 13.6, and the tablet computer group at 18.3, we found that the suitable number of the questions in a quiz was 15.
- In Study 2, different question styles for 10 questions were set, namely multiple choice, fill-in-the-blank, and a combination of both multiple choice and fill-in-the-blank, with the students taking the quizzes on tablet computers. The percentage of fill-in-the-blank was the lowest of the three question styles.
- In Study 2, a combination of both multiple choice and fill-in-the-blank was the best style of the three in terms of increasing knowledge and self-efficacy.
- In Study 3, students took quizzes comprising 15 combinations of both multiple choice and fill-in-the-blank questions. We compared the quiz methods between the "all questions displayed" method and the "series of questions and answers" method. From the results, the percentage of "all questions displayed" method was higher than the "series of questions and answers" method.
- In Study 3, the students who used the "all questions displayed" method recognized high self-efficacy of understanding.
- In Study 4, we focused on the context of quizzes. The percentage of the correct answer was affected by their test approach tendency.
- In Study 4, "low–high" students who were negative towards the tests adapted to a "series of questions and answers" method, rather than all questions being displayed at the same time.

A limitation of this research was that we were not able to investigate students' knowledge about "Introduction to Information Science." Although the students who took part in this research were first-time participants, it is important to analyze the relationship between their previous knowledge and the percentage of correct answers. In addition, both working tasks and collaborative learning are important for students to increase their knowledge and motivation [21–23]. Therefore, we consider that there needs to be a change in the learning style and learning environments, which enable the students to learn from each other.

We consider the following to be important issues to be investigated in the future: (1) to analyze the timing of answers; (2) to compare analysis among question styles using smartphones; (3) to compare analysis between the "all questions displayed" and the "series of questions and answers" methods using tablet computers; (4) to adopt other question styles that we did not adopt in this study; (5) to change the subject about the class; and (6) to do real lessons of practical use.

Acknowledgments This work was supported by JSPS KAKENHI Grant Number 23700979 and 26350310. We thank the Center for Research on Educational Testing (https://www.cret.or.jp/?english) and the participating university students.

Appendix

1. Items of test approach–avoidance tendencies

Test approach tendencies

1. I feel like studying when there is a test.
2. When we have a test, I really want to get a better score than the others.
3. I try to compete positively with other people in tests.
4. I like tests.
5. I like to test my abilities through tests.

Test avoidance tendencies

6. I want to avoid making my lack of ability apparent through tests.
7. I feel anxious about whether I will get a bad score on a test.
8. I don't like being compared to other people through tests.
9. I feel down when there is a test.
10. I think studying is more enjoyable when there are no tests.

2. Items of posttest questionnaire

Study 1

Burden of taking quizzes

1. What do you think about the number of questions on the test carried out using mobile devices?
2. It was a burden to take this quiz using mobile devices.
3. I would prefer taking a quiz using mobile devices more often if there were fewer questions.
4. I would prefer taking a quiz using mobile devices more often if there were more questions.
5. If one mobile-device quiz was given after every class, such as the quiz given this time, how many questions would make you want to take the quiz?

Willingness to have quizzes

6. I was enthusiastic about taking this quiz using mobile devices.
7. This quiz using mobile devices was useful for reviewing the class.

8. This quiz using mobile devices connected what I learned in class with knowledge retention.
9. I want other classes to give quizzes using mobile devices.
10. I would like a quiz using mobile devices after every class.

Self-efficacy

11. I fully understood what I learned in this class.
12. I feel confident about what I learned in this class.
13. The content we learned in this class is an area that I am strong in.
14. I will obtain good quiz scores in this class.
15. I can explain what I learned in this class to other people.

Study 2

Burden of taking quizzes

1. Taking quizzes with only multiple-choice questions is a burden (reversed item).
2. Taking quizzes with only fill-in-the-blank questions is a burden (reversed item).
3. Taking quizzes with both multiple-choice and fill-in-the-blank questions is a burden (reversed item).

Willingness to have quizzes

4. Quizzes with only multiple-choice questions make me answer the questions willingly.
5. Quizzes with only fill-in-the-blank questions make me answer the questions willingly.
6. Quizzes with both multiple-choice questions and fill-in-the-blank questions make me answer the questions willingly.
7. Quizzes with only multiple-choice questions lead to knowledge retention.
8. Quizzes with only fill-in-the-blank questions lead to knowledge retention.
9. Quizzes that combine multiple-choice and fill-in-the-blank questions lead to knowledge retention.
10. Quizzes with only multiple-choice questions are easy to answer.
11. Quizzes with only fill-in-the-blank questions are easy to answer.
12. Quizzes with both multiple-choice questions and fill-in-the-blank questions are easy to answer.

Self-efficacy

13. Quizzes with only multiple-choice questions make me feel that I have sufficiently understood what I have learned in this class.
14. Quizzes with only fill-in-the-blank questions make me feel that I have sufficiently understood what I have learned in this class.
15. Quizzes with both multiple-choice questions and fill-in-the-blank questions make me feel that I have sufficiently understood what I have learned in this class.

16. By taking quizzes with only multiple-choice questions, I have become able to explain to others what I have learned in this class.
17. By taking quizzes with only fill-in-the-blank questions, I have learned to explain to others what I have learned in this class.
18. By taking quizzes with both multiple-choice questions and fill-in-the-blank questions, I have become able to explain to others what I have learned in this class.
19. I sufficiently understood what I learned in this class.
20. I am confident of the content I learned in this class.

Study 3

Burden of taking quizzes

1. It was a burden to take this quiz.
2. I felt there were a lot of questions.
3. I felt there were few questions.
4. It was a burden to take the quiz that had all questions displayed rather than a series of questions and answers.
5. It was a burden to take the quiz that had a series of questions and answers rather than all questions displayed.

Willingness to have quizzes

6. I was enthusiastic about taking this quiz.
7. It was easy to work continually.
8. I am willing to take quizzes using this quiz method.
9. I was enthusiastic about taking the quiz that had all questions displayed rather than a series of questions and answers.
10. I was enthusiastic about taking the quiz that had a series of questions and answers rather than all questions being displayed.
11. The quiz that had all questions displayed was easier to work through continually than a series of questions and answers.
12. The quiz that had the series of questions and answers was easier to work through continually than all questions being displayed.

Self-efficacy

13. I can gain knowledge if I take quizzes every time.
14. I sufficiently understood what I learned in this class.
15. I can explain what I learned in this class to other people.
16. I am confident of the content I learned in this class.
17. I can gain knowledge by taking the quiz that has all questions displayed rather than a series of questions and answers.
18. I can gain knowledge by taking the quiz that has a series of questions and answers rather than all questions being displayed.
19. I understood what I learned in this class enough to take the quiz that has all questions displayed rather than a series of questions and answers.

20. I understood what I learned in this class enough to take the quiz that has a series of questions and answers rather than all questions being displayed.
21. I can explain what I learned in this class to other people so that I am able to take the quiz that has all questions displayed rather than a series of questions and answers.
22. I can explain what I learned in this class to other people so that I am able to take the quiz that has a series of questions and answers rather than all questions being displayed.

References

1. Central Education Council.: Toward the construction of undergraduate education. Retrieved November 13, 2013, (2008) from http://www.mext.go.jp/b_menu/shingi/chukyo/chukyo4/houkoku/080410/001.pdf. Accessed 06/30/2015
2. Kitazawa, T., Nagai, M. and Ueno, J.: Effects of feedback systems in blended learning environments: focus on student satisfaction in information technology education courses. In: Proceedings of the IADIS International Conference on e-Learning, pp. 259–266 (2010)
3. Chan, T-W., Milrad, M., and 15 others.: One-to-one technology-enhanced learning: an opportunity for global research collaboration. Res. Prac. Technol. Enhanced Lear. J. 1(1), 3–29
4. Kuh, G.D.: Guiding principles for creating seamless learning environments for undergraduates. J. Coll Stud. Dev. 37(2), 135–148 (1996)
5. Parsons, D.: Refining current practices in mobile and blended learning: new applications, Information Science Reference
6. Bransford, J.D., Barron, B., Pea, R., Meltzoff, A., Kuhl, P., Bell, P., Stevens, R., Schwartz, D., Vye, N., Reeves, B., Roschelle, J., Sabelli, N.: Foundations and opportunities for an interdisciplinary science of learning. In K. Sawyer (Ed.), The Cambridge handbook of the learning sciences (pp. 19–34). New York: Cambridge University Press (2006)
7. Tabata, Y., Yin, C., Ogata, H. and Yano, Y.: An iPhone quiz system for learning foreign languages. In: 2nd International Asia Conference on Informatics in Control, Automation and Robotics (CAR), 3: 299–302 (2010)
8. Woodill, G.: The mobile learning edge: tools and technologies for developing your teams. McGraw-Hill Education (2010)
9. Cavus, N., Al-Momani, M.M.: Mobile system for flexible education. Comput. Sci. 3, 1475–1479 (2011)
10. Rogers, Kipp D.: Mobile learning devices: essentials for principals, solution tree (2011)
11. Mamat, K., Azmat, F.: Mobile learning application for basic router and switch configuration on android platform. Social Behav. Sci. 90, 235–244 (2013)
12. Quinn, Clark N.: Designing mLearning: tapping into the mobile revolution for organizational performance. Pfeiffer (2001)
13. Sharples, M., Pea, R.: Mobile learning. In: Sawyer, K. (ed.) The Cambridge handbook of the learning sciences, 2nd edn, pp. 501–521. Cambridge University Press, New York, NY (2014)
14. Kitazawa, T.: The design of the test format for tablet computers in blended learning environments: a study of the test approach–avoidance tendency of university students. In: Proceedings of the IADIS International Conference e-Learning 2013, 466–469 (2013)
15. Kitazawa, T., Sato, K.: A comparative analysis of tests using smartphones and tablet computers: perceptions about the number of test questions and motivation for taking the test.

In: Proceedings of World Conference on Educational Multimedia, Hypermedia and Telecommunications 2014 (pp. 807–812). Chesapeake, VA: AACE (2014)

16. Bloom, B.S., Hastings, T.H., Madaus, G.F.: Handbook on formative and summative evaluation of student learning. McGraw-Hill, New York, USA (1971)

17. Koole, Marguerite L.: A model for framing mobile learning. In: Ally, M. (ed.) Mobile learning: transforming the delivery of education and training. AU Press (2009)

18. Suzuki, M.: How learning strategies are affected by the attitude toward tests: Using competence as a moderator. Jpn. J. Res. Test. 7(1), 52–65 (2011)

19. Kitazawa, T., Sato, K., Akahori, K.: The effect of question styles and methods in quizzes using mobile devices on student motivation and percentage of correct answers: focusing on student test approach–avoidance tendencies. Jpn. Soc. Educ. Technol. 38(3), 193–209 (2014)

20. Kitazawa, T.: The effect of question methods in quizzes using smartphones on percentage of correct answers and response time. Res. Rep. JSET Conf. 15(1), 559–564 (2015)

21. Pachler, N., Pimmer, C., Seipold J.: Work-based mobile learning: concepts and cases. Peter Lang Pub Inc (2010)

22. Danaher, P. A., Moriarty, B., Danaher, G.: Mobile learning communities: creating new educational futures. Routledge (2009)

23. Reychav, I., Wu, D.: Mobile collaborative learning: the role of individual learning in groups through text and video content delivery in tablets. Comput. Hum. Behav. 50, 520–534 (2015)

A Generalized Approach for Context-Aware Adaptation in Mobile E-Learning Settings

Tobias Moebert, Raphael Zender and Ulrike Lucke

Abstract Most existing adaptive tools and applications have been developed specifically for one or few selected scenarios. What is missing is a framework that enables the methodical development of adaptive mobile applications that support a wide variety of educational scenarios. However, challenges are to identify relevant contextual information, to enable the design of adaptive learning scenarios, even by non-technophile teachers, to create a platform-independent way of collecting contextual information and to find a way to satisfy the different demands from universities and industry. To meet some of these challenges, this chapter introduces a systematic approach for developing mobile adaptive learning applications.

Keywords E-learning · Mobile learning · Context detection · Context-awareness · Adaptivity · Framework · Mobile devices

Abbreviations

API	Application programming interface
GUI	Graphical user inferface
ITS	Intelligent Tutoring System
MOTIVATE	Mobile Training Via Adaptive Technologies
OWL	Web Ontology Language
RDF	Resource description framework
RDFS	Resource description framework schema
REST	Representational state transfer
SPARQL	SPARQL protocol and RDF query language

T. Moebert (✉) · R. Zender · U. Lucke
Universität Potsdam, Institut für Informatik, Haus 4, August-Bebel-Straße 89,
14482 Potsdam, Germany
e-mail: tobias.moebert@uni-potsdam.de

R. Zender
e-mail: raphael.zender@uni-potsdam.de

U. Lucke
e-mail: ulrike.lucke@uni-potsdam.de

© Springer International Publishing Switzerland 2016
A. Peña-Ayala (ed.), *Mobile, Ubiquitous, and Pervasive Learning*,
Advances in Intelligent Systems and Computing 406,
DOI 10.1007/978-3-319-26518-6_2

1 Introduction

Today, mobile devices like smart phones and tablet computers are affordable and widely used not only in private domains, but also in business applications and educational efforts. The use of these devices for mobile learning made it possible to access educational material spatially and temporally unbounded. This, in turn, has made it possible to use time periods for learning that were unavailable before. Moreover, learning with mobile devices opened up new possibilities for situated learning, i.e. problem-based learning in real-life situations.

However, mobile devices are offering only a reduced screen size and differ in forms of interaction. In addition, users of mobile learning software tend to have a shorter attention span compared to learning in a formal setting [1]. For this reason, the concept of micro learning, with its focus on relatively small learning units and short-term learning activities, is often preferred. To realize the full potential of micro learning, it is important to offer content to the learner that is on the one hand relevant for their current situation and on the other hand customized to their personal needs. In order to achieve this, contextual information can be used to match educational offers with the learner's individual needs and characteristics. All information that describes the interaction between the user, the application and the environment are considered contextual information [2, 3]. Several research initiatives have identified a plethora of contextual information that might be used for adaptation. Mobile devices offer a variety of sensors and interaction options that can be used to gather contextual information. The challenge is to identify those pieces of contextual information that are considered most relevant by the authors of educational software and can still be reliably detected.

Most existing adaptive tools and applications have been developed specifically for one or few selected scenarios, resulting in some rather specialized prototypes. What yet is missing is a framework that enables the methodical development of adaptive mobile applications that support a wide variety of educational scenarios. Such scenarios could originate from academic as well as business areas. An exemplifying use case for an academic is a biological excursion during which contextual information could be used to lead the students to interesting plants or animals, to help them identify information that is relevant for their learning goals, or to match content with their personal likings.

This would be even more interesting if it was a multi-day excursion. According to [4], a more business-oriented use case would be that of a relief organization that specializes in helping refugees and other people in need. Such an organization operates worldwide and is therefore dependent on a large number of employees as well as volunteers. To train these people, the organization already relies on class lecture, classic e-learning and mobile learning. Mobile learning is especially crucial when it comes to delivering training courses to people who do not have access to class lectures or classical e-learning means. With adaptation mechanisms, the mobile learning system would in future be able to automatically adjust to ambient conditions (e.g. location, noise level, but also temperature and humidity) and to leverage parameters like the user's experience to accurately deliver a suitable learning unit, in an adequate form of media.

In the course of this chapter we want to present a framework for the systematic creation of adaptive mobile applications (Sect. 2) that we developed. We have placed great emphasis on application in both academic and industrial areas, a systematic detection of relevant contextual information (Sect. 3), platform-independence (Sect. 4.2) and finally finding new user experience concepts that ease the use of adaptive applications for authors as well as end users (Sect. 4.3).

2 Adaptive Learning

Since the development of the first teaching machines in 1924, computer-aided learning without a teacher's presence is an important topic in educational research. First attempts followed the theory of behaviorism and simply rated a learner's answers to specific questions [5]. Later, cognitivism-driven research identifies phases of human learning and aimed at their support with the help of intelligent tutoring systems (ITS) [6].

In general, ITS utilize a set of models about domains, learners, and didactical strategies to derive decisions regarding learning process support. The central student model contains the system's beliefs about the learner's knowledge, derived by sources like the history of the learner's behavior [7]. In the last 10–15 years, the trend of mobile devices with a broad spectrum of sensors as well as a permanent internet connection led to a broad spectrum of measurable information about a learner's physical context, social relations and even physical functions, like the heartbeat rate. Educational systems are therefor able to draw on abundant context resources to adapt to learner's as well as teacher's needs, and increase the convenience and efficiency of education in a proactive, pervasive manner [8].

Constructivist learning scenarios in terms of contextualized learning became technically realizable. They connect abstract learning content to specific use cases and, in particular, to relevant physical situations (situated learning) [9]. This leads to the requirement of adaptive IT solutions to gather, transform, connect, and interpret contextual data as well as react to the learner's current needs in a didactically sensible manner.

2.1 Mobile Adaptive Learning Systems

Adaptive learning systems make use of context information to match the form of presentation, structure, and selection of learning content to a learner's individual preferences, previous knowledge, and learning goals. Two trends of the last years led towards a new generation of adaptive learning systems: Powerful mobile devices and an increased flexibility for the selection and creation of e-learning material.

On the one hand, the access to educational offers by individual, mobile devices with high performance, multiple sensors, as well as a new level of integration into the learner's everyday life allows a broad context acquisition, far beyond the possibilities of previous device classes. Today's smart phones and wearables measure location, orientation, movement, and even physical functions of their users. They have access to calendars, contact data, personal documents, and several other data concerning the organization and social aspects of a learner's everyday life.

On the other hand, these devices create a need for suitable content. Learners are able to access all kinds of information anytime and may choose that piece of information which is most suitable for their current situation. Fortunately, document standards and interoperability concepts like responsive websites allow teachers to create content for different target platforms. Dedicated authoring systems that are partly integrated into learning platforms like, Moodle enable even beginners to make use of these technical possibilities. However, content is hard to present on a device with limited expressive capabilities and to a learner with a limited attention (e.g. during a train ride). Thus, from a didactical perspective, the concept of micro-learning [1] is often preferred in mobile learning arrangements: Small and self-contained learning units that can be worked through easily, even on the move.

The combination of mobile devices and flexible micro-learning content results in adaptivity mechanisms for mobile devices from automated selection of content and context-aware presentation as well as the adaption of learning paths to individual learning styles and goals. Even flexible and adaptive didactic settings for different situations are included [9]. A *mobile adaptive learning system* is an interactive system that personalizes and adjusts e-learning content, pedagogical models, and interactions to meet the individual needs and preferences of users if and when they arise [10], with a special focus on mobile end user devices for context acquisition.

Following the approach of micro-learning, the learning objects and processes become decoupled from the time and place of learning. Yet, the adjustment of educational offers to the given environment, activity, etc. is just a common motivation and benefit of using mobile devices for learning. Detecting the relevant context information (like location, current task or need, surrounding objects and persons, individual learning style and progress, etc.) and analyzing it towards a proper adaptation of a learning offer is a major characteristic of mobile adaptive learning.

2.2   Related Work

A broad spectrum of related work focuses on visions regarding the potential of context-aware systems for education or the utilization of context information for specific learning scenarios [9]. Most of the research in this field can be seen as non-mobile and stand-alone solutions for specific scenarios (e.g. for lecture recordings [11], ITS [12]) or is not suitable for mobile learning due to its complexity [13].

The following related work [14] gives an overview about research in the field of mobile adaptive learning systems. Five pedagogical settings have been identified [15]. The majority of existing mobile context-aware e-learning applications can be classified into:

- *Formalized settings* that take place within the bounds of a dedicated educational institution.
- *Physical settings* that take place in an authentic physical context (e.g. field trips).
- *Immersive artificial* settings where the learning experience is augmented by virtual artefacts, simulations, or pervasive educational games.
- *Collaborative settings* that integrate learners into communities, where collaboration can be supported by recommendation or awareness tools, among others.
- *Loose settings* that take place independent of the current context.

Many existing mobile tools and applications can be assigned to one of the first four settings. They have been developed specifically for selected scenarios, resulting in some rather specialized prototypes, e.g. tangible tabletops [16], personal response systems [17], tools for language learning [18], field trip support [19], museum guides [20], collaborative editing [21], and so forth.

Other research focuses on a more generic approach to educational context-awareness and tries to capture all kinds of context information, including the learner's context, in order to bring forth adaptive frameworks for seamless, personalized learning. For instance, a rule-based system for the selection and presentation of learning content that incorporates a reward/penalize mechanism for further refinement by feedback has been proposed [22]. However, the main focus of this system effectively lies on technical context, such as device specifications, and corresponding adaptation rules for adaptive rendering of learning content.

A similar framework [23] targets an adaptive engine that uses information about learning styles as well as context-awareness in order to provide appropriate, personalized content. What all of these endeavors have in common, though, is that they concentrate on mobile learning in loose settings. So far, no frameworks can be reported yet that encompasses the whole bandwidth of possible didactical settings.

Likewise, the state of research is lacking in analysis of the relation between learning settings and context information in terms of different types, relevance for the setting, availability of the context information, and its reliability. Furthermore, least approaches take into account the teacher's role in authoring e-learning content. However, these considerations are essential to successfully transfer results of e-learning research into current educational development beyond academic visions.

2.3   Roles and Perspectives

Beside technological possibilities as well as limits, specific mobile adaptive learning systems are affected by and the result of an interdependence between authors of learning material (teacher) and mobile users (learner).

Authors of learning material are often employed by educational institutions or by commercial providers of educational material. In particular, teachers come with different levels of IT knowledge. It cannot be assumed that they understand the possibilities and mechanisms of adaptive mobile learning at all. Teaching with adaptive software is a concept that is quite new for teachers and educators nowadays, because of the lack of practical implementations (in the field of mobile adaptive software) that are really used in educational routine [14]. Thus, a suitable adaptive mobile learning system must hide the technical and didactical complexity of adaptation, but at the same time provide tools to connect learning content to contextual information.

Mobile learners are users of adaptive mobile learning systems in terms of consuming and interacting with the learning content, created by authors. They may also be non-technophile and not familiar with mobile adaptation of learning content. While teachers have to be supported in their selection of context and the connection to content, learners may not be aware that their context is used to adjust their learning experience to their current needs. Therefore, the adaptation processes must be invisible to simplify learning on the one hand. On the other hand, unexpected adaptations may irritate learners as they feel they lose control over their devices. This irritation must be avoided by a highly intuitive user interface with a suitable transparency of background adaptation processes. The first step to a user friendly systematic adaptation for teachers and learners is the in-depth analysis of relevant context data for mobile and micro-learning settings.

3   Context

Weiser [24] coined the term *ubiquitous computing* as a concept where computing can occur everywhere and anywhere using any device in any location. From this term the concept of context-awareness was introduced [25]. Context aware devices therefore are trying to make assumptions about the user's current situation and are trying to present services and information based on these assumptions. It is believed that contextual information can be used by mobile learning software to adapt its content to the preferences, needs, knowledge and learning objectives of its user [26].

One must distinguish, however, between context and situation, as context describes just the types of relevant information whereas situation describes the actual values for these types [27]. Detecting the relevant context information (like location, current task or need, surrounding objects and persons, individual learning style and progress, etc.) and analyzing it towards a proper adaptation of a learning offer is a major characteristic of mobile, pervasive learning approaches. Related work [3] compiled an extensive collection of context information that might be relevant for adaptive learning software. Open, however, is the question of what contextual information by teachers are considered to be relevant for creating adaptive learning content. To find out which kinds of information are considered to be relevant by teachers, we decided to analyze different existing educational scenarios.

3.1 Gathering of Educational Scenarios

To have a basis that is as varied as possible and to respect the project's overall goal to develop a framework that may be used in an academically as well as in a business setting, different educational scenarios from both areas needed to be considered. To achieve this, the educational scenarios had to be gathered by a university among teachers and educators as well as adjusted by a business provider of professional learning software. The business provider collected own scenarios among their customers and introduced them into the selection of relevant context information. In order to have these scenarios described in a semi-structured manner, an educational scenario form sheet was developed consisting of:

- an optional graphical representation of the learning scenario itself in form of a Didactic Process Map
- a short description of the scenario's general setting
- a list of desired educational objectives
- a detailed sequence description

In total, we collected 13 distinct scenarios from local teachers of science, humanities, economics and pedagogies as well as 8 scenarios from business and industry. Using these scenarios we evaluated a list of possibly relevant contextual information. An existing classification [27] divides context into the four basic classes of technical, physical, personal and situational context. Because the project has a strong focus on mobile devices, we decided that the mobile part of the physical context had to be considered separately. In a similar fashion, we considered the scenario part of the situational context separately to take into account the scenario-driven analysis approach that we had chosen.

3.2 Identification of Relevant Contextual Information

From the gathered educational scenarios we derived the following subset of context information and sorted them into the aforementioned context classes:

- *Physical context*: weather (current and future), ambient noise, luminosity, time, humidity, temperature
- *Mobile*: location (separated into address, building, region and country), arrival, departure, distance to a point-of-interest, means of transportation, speed, destination
- *Situational context*: body gesture, facial expression, viewing direction
- *Scenario*: learning progress, current task, time required for processing
- *Personal context*: appointments, prior-knowledge (verified and derived), motivation, expectations and motifs, preferences, social relations, disabilities
- *Technical context*: available infrastructure (like printer and external monitor), device capabilities (like voice recording, image capturing, video recording and audio output)

To make the different educational scenarios comparable, we classified them into the five educational settings [15]. Furthermore, we created a questionnaire that was distributed to the contributors of each educational scenario to find out which kinds of contextual information is considered relevant for educational content adaptation by actual teachers, educators and business professionals. The feedback gathered was evaluated and analyzed with respect to relevance. Information from the following context classes has been estimated to be relevant for the defined settings:

- *Formalized Setting*: no significance for any of the context classes.
- *Physical Setting*: Personal, Scenario.
- *Collaborative Setting*: Personal, Scenario.
- *Immersive Setting*: Physical, Personal, Situational, Scenario.
- *Teaching and learning support*: Personal, Scenario.

Additionally, we questioned experts in mobile learning software on how they would estimate the accuracy of detection by today's means, aggregated over the specific educational scenarios. The information from the following context classes was estimated to be accurately measurable:

- *Formalized setting*: Scenario, Technical.
- *Physical setting*: Physical, Mobile, Scenario, Technical.
- *Collaborative setting*: Scenario, Technical.
- *Immersive setting*: Physical, Mobile, Personal, Situational, Scenario, Technical.
- *Teaching and learning setting*: Physical, Scenario, Technical.

Figure 1 provides a graphical representation of the findings from the combination of our data on both relevance and measuring accuracy. The chart is divided into four quadrants of which each one can be described as containing educational settings with context information that is considered to be either:

- *First quadrant*: relevant and accurately measurable.
- *Second quadrant*: relevant and inaccurately measurable.
- *Third quadrant*: irrelevant and inaccurately measurable.
- *A Fourth quadrant*: irrelevant and accurately measurable.

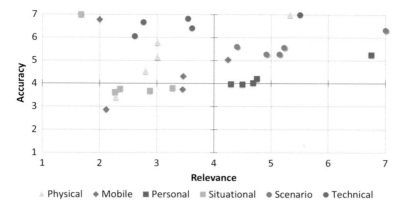

Fig. 1 Combination of the collected data on relevance and measuring accuracy

It is apparent that for further research, context classes that contain information that have been estimated to be either very relevant, very accurately measurable or qualify in both aspects are of utmost interest.

- *Relevant*: Scenario, Personal, Physical (partly), Technical (partly).
- *Accurate*: Scenario, Technical, Physical, Mobile (partly), Personal (partly).

3.3   Implications from the Analysis

Our collected data indicate that the considered information from the scenario context (e.g. learning progress, current task) and the personal context (e.g. prior-knowledge, motivation) have been estimated to be relevant in most of the educational settings. In these settings, information belonging to the scenario context provides an overall good measuring accuracy, and information belonging to the personal context can be measured with at least tolerable inaccuracies.

Most interesting is the fact that information originating from the mobile context (e.g. location, speed, destination), which can be gathered quite accurately with current mobile devices and is often the first choice when it comes to mobile content adaptation, has been estimated to be predominantly irrelevant for individual educational context adaptation. Similar statements can be made for information belonging to the technical and physical context class (except for immersive settings).

This discrepancy can have many reasons. The most obvious could be that this information is indeed irrelevant for educational context adaptation (at least for the moment) and has just been overvalued because it is the most obvious choice when it comes to mobile learning and mobile devices. Another reason could be that teaching with adaptive software is a concept that might be quite hard to grasp for teachers and educators nowadays, because of the lack of practical implementations that have found their way into everyday school life. In the end, research might be facing a so-called "chicken or the egg dilemma" here. On the one hand, it is hard for people to think of features they would like to see in software which is mostly unknown to them or even has an alienating effect. On the other hand, it is difficult to develop useful software concepts when the specific requirements are unclear.

4 Framework

This section deals with the framework proposal for the creation of adaptive mobile learning applications. A closer look will be taken on the general architecture, platform-independent context detection as well as concepts for a dedicated mobile user experience.

4.1 Adaptation of Learning Content

In order to adapt learning content to suit learners' needs and preferences as well as their prior knowledge and learning objectives, adaptive learning systems make use of several detailed models:

- *Content or domain models* describe what is to be learned within a course or discipline. Causal relations between learning units are modeled as well as distinct levels of difficulty or (methodological) forms of learning.
- *Learner models* keep track of a learner's progress, that is, their remaining deficiencies with respect to the desired level of knowledge, including mistakes made while solving tasks. Required actions can be derived from these sources of information.
- *Didactic or adaptation models* encompass possible forms of reaction to tackle remaining gaps in knowledge, depending on the given need for action. So far, these reactions are defined only in a rather static form, though.

This schema is inspired by Intelligent Tutorial Systems [12] which combine approaches from cognitive psychology and artificial intelligence. Under the designation adaptive systems, this area of research currently receives an increasing level of attention, factoring in social and pedagogical aspects, in particular. The adaptation of the didactic approach to the learner's current progress and needs can even be used more effectively by further including information about the learner's environment and current situation, so-called context. Information about e.g. location, physical parameters, like brightness or noise level, motion, direction, or persons and objects in the learner's vicinity can be leveraged to a more personalized, while at the same time socially embedded, and thus, more successful learning experience.

A personalized learning experience is created for the learner in several steps. Their needs and preferences are included as well as their current context. Our project's conceptual and technical novelty consists in a multi-staged process, which has not been known from existing approaches and which contributes significantly to the solution's suitability for mobile systems and in particular to its re-usability.

In a first step, an adaptable (i.e. manually adjustable via the variation of certain parameters) application for learning is designed with the help of an authoring system which is to be part of the developed framework. In a second step, this application is deployed onto the learner's mobile device where it serves as an adaptive (i.e. automatically adapting to a given situation) application for learning. Its specific form results from a semi- or fully automatic adaptation to the learner. The core of our solution is the framework for the creation of those learning applications. It will provide a means to generate applications for mobile devices which are then adaptable under certain aspects. This is accomplished on a higher level by an adaptation engine, which again accesses several components:

- *Adaptation rules*: In the relief organization scenario, for instance, a module for vehicle control within a convoy may be shown only to those learners who have a driver's license. This can be modeled by adaptation rules.
- *Learning content and audio-visual material*: To these belong e.g. specific video or audio instructions and explanations for vehicle control in the relief organization scenario.
- *Context detection*: In the above-mentioned scenario, relevant context information might be the specific type of vehicle or the place of action, in order to provide specific content for learning about traffic law in the region.
- *Learning scenario*: Within the learning scenario, the order in which selected learning units may be recommended can be modeled, for instance. Referring to the running example of the relief organization, a general training unit about traffic law could be placed before the more specific training unit about vehicle control. These components can be added or removed in a modular fashion with the help of an authoring system.

4.2 Architecture

Figure 2 depicts a draft of an exemplary implementation of the proposed framework. The system will create a personalized learning experience in a multi-step process and is divided into the three main components authoring system, rule generator and mobile application. The authoring system is used here for teachers to create and manage adaptable (i.e. manually adjustable by variation of parameters) micro-learning content. The created learning contents are transferred to the mobile learning application where they are available as adaptive (i.e. automatically adjustable) learning content.

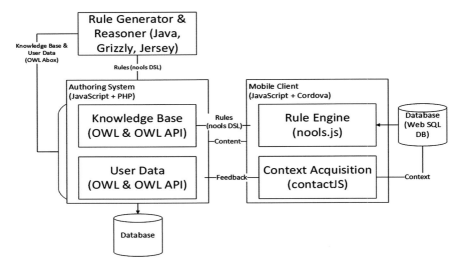

Fig. 2 The architecture of the framework with its server side components on the *left side* and the client side components on the *right side*

In order to implement the adaptivity of the mobile learning application, conclusions must be drawn both from the user behavior as well as the collected context information. For this purpose, a reasoning mechanism is necessary which, however, cannot run on the mobile device itself due to the limited resources of mobile devices and the lack of a platform-independent reasoning framework. Outsourcing the reasoning on a server to only trigger it when necessary is not possible either because it has been identified as an essential requirement of the application to be usable without an internet connection.

To meet these requirements, a hybrid approach has been chosen. Thus, the reasoning is indeed outsourced to a server, but is triggered by the rule generator just before deployment to the mobile device. This uses both the knowledge base and possibly collected feedback about the behavior of the learner and creates a set of rules. The rule set is evaluated by a rules engine in the mobile learning application to dynamically select or adjust learning contents. The inference of high-level contextual information takes place on the mobile device as part of the context acquisition. Figure 3 shows a simplified representation of the data flow.

A variety of programming languages, frameworks and technologies will be used hereby:

- *Java* will be mainly used for the rule generator. In combination with frameworks like Project Grizzly and Jersey, it has been used to build a RESTful web service that provides interfaces for rule generation and information about different parts of the adaptation engine or context detection (e.g. supported contextual information).

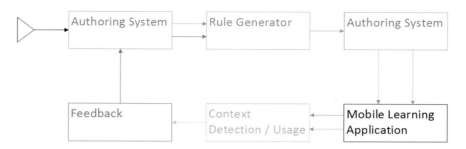

Fig. 3 The data flow of the complete infrastructure. Rules at first are solely generated from the content created by an author. In a later step, collected information about the user's behavior (feedback) can be used to refine the originally generated rules

- *JavaScript* is used in many different parts of the whole project. On the server side, the authoring system leverages JavaScript to create the user experience. On the mobile client side, JavaScript is used for the context detection (see Sect. 4.2).
- *OWL* the Web Ontology Language is a specification of the World Wide Web Consortium to create, publish and distribute ontologies using a formal description language. It is based on RDFS and part of the W3C Semantic Web Activity. OWL is used in combination with the OWL API to store the knowledge base and the collected user behavior as well as to draw conclusions from these data (reasoning) for adaptivity.
- *SPARQL* is a query language for RDF. It is used, among others, to combine information about the user's behavior and the collected contextual information.
- *Nools* is a rules engine based on the Rete algorithm written entirely in JavaScript. It is used to evaluate the rules that were created by the rule generator and accordingly give proposals for adaptations.
- *Apache Cordova* is a set of device APIs that allow a mobile app developer to access native device function such as the camera or accelerometer from JavaScript. It is used in combination with the context detection and rule engine to create the adaptive mobile learning application infrastructure.[1]

4.3 ContactJS[2]

Modern mobile devices like smart phones or tablet computers, that are already being used to consume mobile learning content, are equipped with a variety of

---

[1]https://cordova.apache.org/.

[2]The work in this part of the project was implemented as part of a master's thesis. Stefanie Lemcke, who worked on the master's thesis, has been supported in her work by the *Stiftung Industrieforschung* with a scholarship.

sensors and interaction options that can be used to gather contextual information. When it comes to the detection of such information, a challenge, however, is to address the heterogeneous landscape of devices and operating systems (and thus programming languages) that learners are using.

Currently, there are many different smart phone and tablet computer developers in the market each using their own operating systems (e.g. iOS,[3] Android,[4] and Windows Phone[5]) and programming languages (e.g. Objective-C, Java, and C#). A context detection system should support all or at least the most commonly used platforms. Development and maintenance of such a system individually for each programming language would be an expansive task and almost impossible to afford by individual educational institutions.

Fortunately, modern mobile devices are supporting the concept of so-called web applications. Such applications are using technologies like HTML, CSS and JavaScript normally used for the development of web sites. In combination with frameworks like Cordova it is possible to create mobile applications that can be ported to different operating systems at reasonable expense.

There is a range of existing frameworks that can be used for context detection like Context Toolkit [28], Citron [29], CASanDRA [30], LoCCAM [31], CASS [32], Hydrogen [33], Gaia [34], Context Fabric [35], CoBrA [36], Solar [37], SOCAM [38] and JCAF [39], to name a few. Most of these frameworks are based on a client-server model with distributed context detection. Solely LoCCAM offers an infrastructure for context detection on mobile devices but was only developed to be used on devices running the Android operating system. Moreover, only Context Toolkit, Context Fabric, CoBrA, LoCCAM and JCAF had usable source code available so that the remaining frameworks where excluded from further consideration.

Due to our focus on context detection and the fact that CoBrA and Context Fabric are also offering inseparable reasoning mechanisms, that exceed the targeted functionality and are too complex to port, we decided to focus on Context Toolkit, LoCCAM and JCAF as potential candidates for a port.

- *Loosely Coupled Context Acquisition Middleware* (*LoCCAM*) is an infrastructure for context acquisition for devices that are running the Android operating system. Its main focus is to use smart phones as the main element for context detection. This may include local and remote resources as sources of information. The systems context detecting components are self-regulating which is achieved through the use of the OSGi framework as the runtime environment. The detection and processing of actual contextual information is realized through Context Acquisition Components (CAC). The collected contextual information is managed and published with the use of System Support for Ubiquity (SysSU[6]).

---

[3]http://www.apple.com/ios/.

[4]http://www.android.com/.

[5]https://www.windowsphone.com/.

[6]https://code.google.com/syssu/.

- *Java Context Awareness Framework* (*JCAF*) is a Java-based framework for context detection and publishing. It is divided into two parts: the Context-awareness Runtime Infrastructure and the Context-awareness Programming Framework. The first part provides the developer with a basic frame for developing a context-aware infrastructure. It is designed as a service-oriented architecture and communication is realized through Java RMI. The second part is an API that provides interfaces for context-aware applications to communicate with the framework. Both parts of the framework are together implemented as three-layer architecture. The Content Client Layer acts as the API, the Content Service Layer manages entities as well as context transformation and aggregation, and the Context Sensor and Actuator Layer task is to actually gather contextual information or to manipulate the context.
- *Context Toolkit* is a framework developed for the use of distributed context detection hardware [28] and can be seen as one of the first frameworks developed for that purpose. Its main focus is on context detection and processing. The framework primarily consists of the five components Widget, Service, Aggregator, Interpreter and Discoverer. Widgets and Interpreter thereby have the tasks of context detection and processing, Services are used to execute orders and actions, Aggregators are used to combine different data for a single entity and Interpreters are used to format and interpret the collected data. The components are realized as abstract Java classes that provide all the basic functionality where concrete features need to be implemented by the developer.

Before we could select a framework to be ported some functional and non-functional requirements for the resulting system had to be noted. Functional requirements are:

- *Context detection*: The system needs to be able to detect contextual information that can be used by learning software to adjust its content to the preferences, needs, knowledge and learning objectives of its user. Sources for that information should be sensors as well as software components.
- *Context persistence*: There is a need for a structure to save gathered contextual information either for a short time or long periods of time.
- *Context processing*: Initial processing steps like interpretation or transformation of the gathered data should be possible. This should be limited to simple conversions like changing the format of the data or for example retrieving the address for geographic coordinates. Complex abstraction work will be done by the hosting application.
- *Context publishing*: For context-aware applications to react to gathered contextual information, it is important to be able to query certain information. This should be possible on demand or if needed in a callback-like manner.
- *Communication*: The interaction between the components of the framework needs to be adapted for the use on mobile devices. It is important to be able to interact with existing hardware and software components of the device.
- *User context management* (*optional*): To protect the user's privacy, it should be possible to inspect and manage the gathered contextual information. This means

that the user would be able to delete information and to add or remove certain context sources. Because context aware applications rely heavily on contextual information it could be problematic if a user refused to reveal required information. This deliberation took us to the conclusion that the decision of which information to gather and which not is in responsibility of the hosting application and that such management on the detection layer can be seen as optional.

The non-functional requirements are:

- *Platform independence*: A main goal of this project is to create a context detection system that can be used on a wide variety of platforms.
- *Operability and maintenance*: The resulting system is intended to be a basic framework for context detection. It will provide an environment to detect and publish contextual information where concrete implementations like the connection of a sensor need to be done by the user.
- *Flexibility*: Supporting a wide range of context sources from both software and hardware is desirable. This includes a simple but open structure for data persistence.

Table 1 shows an overview of the most important features for each of the frameworks. Each framework focuses on the detection and publishing of context information. Thus, independence between the collection of context data and the usage of such data by a context-aware application is given. Despite this similarity each of the frameworks has its own strong points. LoCCAM already focuses on mobile devices (Android only, though) and uses a self-regulation mechanism to start and stop context detection components. JCAF uses a service oriented infrastructure for context detection and publishing and focuses on distributed systems. Context Toolkit uses a concept similar to JCAF but provides separate modules for context processing and registration and is free of any dependencies. Ultimately, the decision was made in favor of Context Toolkit. Mostly because it is the only framework that had no dependencies that would have been needed to port as well.

As previously stated, our goal was to port an existing context detection framework to JavaScript. A strong focus was set on platform independence to support as many mobile devices as possible. However, two questions needed to be answered first before a port could be done.

Table 1 Prominent features of the port candidates

Feature	Framework		
	LoCCAM	JCAF	Context toolkit
Language	Java	Java	Java
Application	Mobile devices (android)	Distributed systems	Distributed systems
Modules	CAC (detection and processing)	Monitor (detection) services (processing) actuators	Widget (detection) service (actuator) aggregator interpreter
Dependencies	OSGi, SysSU	Java RMI	

The first question that arose was which implementation variant to choose. On the one hand, the framework could have been implemented as a standalone application which would run on the target device. The advantage would be that only one context detection system would be running on the device providing contextual information to all context aware applications. A serious disadvantage would be that each mobile operating system has another way of distributing data between running applications. A context detection application would need to take each of these into account and thus would limit its platform independence.

On the other hand, an implementation as a library was conceivable. The library would be included by future context-aware applications and provide all the functionality needed to gather contextual information. Such a solution could be designed to work around the problem of distributing the gathered contextual information between individual applications because each application that used the library would have direct access to the gathered data. The main disadvantage of such an implementation would be that different applications using the library would gather and store their own data and so would inevitably produce redundant data. Eventually, the decision fell in favor of to the library solution because the number of concurrent e-learning applications can be expected to be low and it would bypass the data distribution problem, yet could be used to implement a central context detection application later.

The second question that had to be answered concerned the components to be adopted. Basically, we planned to take over the existing structures and adapt them only to the selected technology. It turned out, however, that some components and concepts conceived to realize the communication in a distributed system would be unnecessary or needlessly complicated in a system where all parts are situated on the same device. With that in mind, the following components were selected for porting:

- *Widgets* are the main components of the Context Toolkit. Their task is to gather contextual information and they are therefore directly connected to context sources such as sensors. They were fully ported, including the structures for the modeling of the detected data and the control of the communication between the components.
- *Aggregators* are consolidating context information about a single entity like a person, a location or an adaptation rule. Hence, they are context gathering components, as well. Contrary to Widgets, they are not directly connected to a context source but rather are collecting different information that is provided by Widgets and Interpreters. Aggregators were ported closely to the original system. Again, the main task will be the consolidation of contextual information. Furthermore, Aggregators will manage the communication with external applications that are using the framework as well as the persistence of the gathered information.
- *Interpreters* process information to achieve format conversions or to abstract low-level context into high-level context. They can be accessed by Widgets or

Aggregators and the complexity of the abstraction process is to be chosen by the programmer.

- *The Discoverer* acts as a registry and provides an overview of all available Widgets, Aggregators and Interpreters. The Context Toolkit allows the existence of multiple discoverers. This was necessary because in a distributed system, parts of the network could be safeguarded, which might prevent communication with a global discovering instance. However, on a mobile device where a discovering service would be accessible to every other component, such a problem would not arise and thus the system can rely on a single Discoverer. In addition to the features defined in the original framework, the ported version covers the communication between different components.

The following components were omitted:

- *The BaseObject* manages the details of the communication in a distributed system. These details include the underlying transport protocol and the exchange format used to transport data through the network. In a local system, all components communicate directly and thus the use of the BaseObject component is obsolete.
- *The DataObject* encapsulates every bit of data that is sent between the components of the system into a general data structure and hereby prepares the data to be transformed into XML and accepts the data retrieved from parsed XML, respectively. This covers requests to as well as responses returning from a component. Again, in a local system such a transformation is not necessary, because all components communicate directly.
- *Services* in Context Toolkit act as actuators and are used to change or influence the environment. As this is not a requirement of the project, the component was not ported but could possibly be added later.

For the implementation of the proposed JavaScript library, a few existing frameworks were used. These are:

- *RequireJS* was used as a file and module loader for JavaScript. Its main benefit is that JavaScript files can be easily managed in a modular way similar to Java-Imports. Additionally, required JavaScript files are loaded on demand. This results in cleaner code and improved processing speed.[7]
- *ease.js* brings object-oriented concepts like classical inheritance, abstract classes and methods, interfaces and static and constant members to JavaScript. It was used to ease the port from the class-based programming language Java to the classless script language JavaScript.[8]
- *r.js* is an optimizing tool as a part of RequireJS. It can be used to minimize, optimize and combine different modules into a single JavaScript file. It was used to generate the final single library file.

---

[7]http://requirejs.org/.

[8]https://www.gnu.org/software/easejs/.

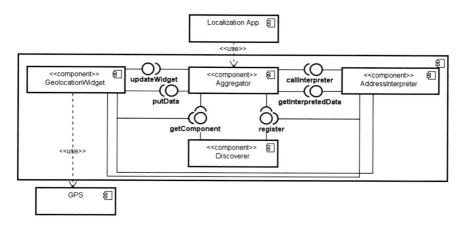

Fig. 4 An example configuration of an application using contactJS to retrieve the address of the user's current location

- *Math.uuid*[9] offers functionality to generate Universal Unique Identifiers (UUID) in JavaScript. UUIDs are used in Widgets, Aggregators and Interpreters and are required for the registration and search by the Discoverer. The library supports the creation of UUIDs following the RFC 4122 Version 4 [40].

The implementation follows the concept that was explained above. The most important changes, compared to the original system, have been made in the area of communication. The components solely required for the communication in distributed systems have been discarded and the concept has been adapted to the communication on local systems.

The Discoverer is now a central component of the framework. Every other component has to register itself with the Discoverer, and future communication is established by acquiring required all components from the discoverer through the use of UUIDs. Widgets and Aggregators have also undergone some changes. Widgets don not save a history of past contextual information anymore; this is now done by the Aggregator. Aggregators are now also responsible for the communication with applications that are using the framework.

Figure 4 shows an example configuration of an application using contactJS. In this example, the application is using the framework to acquire the address of the current position of the user, which might be relevant on a field trip. A widget was deployed to acquire the position of the user directly from the GPS sensor of the mobile device and an interpreter encapsulates the reverse geocoding functionality provided by an external source (e.g. Google or OpenStreetMap).

---

[9]http://www.broofa.com/Tools/Math.uuid.js.

Both components are registered with the discoverer and are publishing their information to the aggregator which in turn communicates with the hosting application. Similarly, other types of context can be managed.

4.4 Mobile User Experience

As mentioned earlier, adaptive mobile learning applications are able to detect the user's situation by evaluating contextual information and using this information to select and adjusting content accordingly. Typically, this will happen in the background unnoticed by the user.

In Sect. 3, we have already shown that the information used for the adaptation processes is often limited in terms of accuracy and reliability. This can lead to the application behaving in a way that is no longer understandable or even alienating for the user. Moreover, the detection of contextual information can pose the risk of violating the user's privacy.

To prevent these problems, the collected contextual information and the adjustments provided on the basis of this information must be transparent for the user. The challenge hereby is to adjust the flow of information in such a way that the user will not get overloaded with information, and thereby deflected from the actual learning task, yet will not be under-served with information, which would create the aforementioned problems.

In the course of the project, we tried to identify user experience design patterns that tackle the problem of information supply and privacy violation caused by a possible lack of transparency. At the beginning of the development of the framework's user experience component, the functionality to be addressed was being determined from nominal patterns of adaptive user experience. Those patterns were elaborated from various user-approved examples.

A digital paper-prototype of an example scenario for adaptive mobile applications was created and made a representative study on the implemented principles possible. It features the following three key techniques:

- *Common user interface components*: Consist of components chosen from the platform's default UI set. Conforming to the platform's interface guidelines, these components are widely proven.
- *Real-life context simulation*: Simulated adaptations based on context changes that happen to the most mobile users every day.
- *Example scenario*: Features a real content scenario that is already adopted in existing applications.

Hereinafter, the user experience patterns that were used or developed, shall be explained in more detail.

On-boarding. As seen in many current mobile applications, an on-boarding process is used to make users familiar with interaction patterns that they would not have been exposed to on a regular basis. The on-boarding is often a set of different

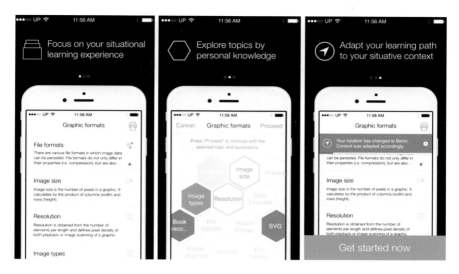

Fig. 5 The on-boarding views that are highlighting our focus on situational learning, the explorative honeycomb navigation view and the way newly detected contextual information is displayed

informational screens shown to the user on the first application start. They provide a brief introduction to the application and highlight and explain the key features. Again, the goal of the on-boarding screens is to make the user familiar with the things he does not see in most of the applications he is already using. In this case we decided to highlight our focus on situational learning, the explorative honeycomb navigation view (see Fig. 5), and the way newly detected contextual information is displayed.

Tutorial overlay. Another user experience pattern that is often used to introduce users to certain app features are so-called coach marks. They act as an indicator for the actions that can be performed on the application and explain what the controls on the application are meant for. We decided to use these coach marks in form of a tutorial overlay screen, a black half transparent screen that lies on top of the actual controls and is presented at the first application start.

Learning content navigation. Former research [41] identified that learning and teaching styles differ in the way learning content is presented to and grasped by the learner. A distinction is made among others between a sequential and global learning and teaching style. Sequential learning hereby "[...] involves the presentation of material in a logically ordered progression, [...] When a body of material has been covered the students are tested on their mastery and then move to the next stage." To support this classical way of learning we developed the *Learning card stream* (see Fig. 6).

Every card in the stream represents a micro learning unit. New cards are added on top of the stream when a card is consumed or contextual information changes.

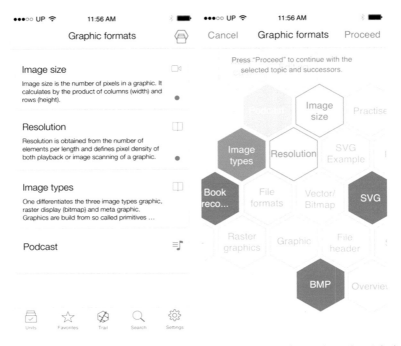

Fig. 6 On the *left* is the card stream to support sequential learning and on the *right* is the honeycomb view that aims at supporting global learners

New cards are flagged to be easily recognizable. To give the learner a general idea of what content might await him, every card has a title, a short description and an icon that will hint at the content type. Though a linear learning style is in focus here, a degree of freedom is given as older cards can be explored through backtracking.

In contrast to a sequential learning style where learners "[…] follow linear reasoning processes when solving problems […] learn best when material is presented in a steady progression of complexity and difficulty" [41] learners who follow a more global style of learning tend to "[…] learn in fits and starts […] make intuitive leaps […] sometimes do better by jumping directly to more complex and difficult material." [41]. To support such a global learning style, we developed the honeycomb navigation view.

All learning units of a scenario are arranged as a honeycomb structure where related learning units are positioned in immediate vicinity. This enables the learner to take a glance at future learning units. Currently reachable learning units are colored blue. Selecting such a learning unit will display direct follow-up learning units, another way to estimate the following educational trail. Which learning units will follow is based on contextual information and the relations between learning units that were defined by the author. We hope that this will help the learner to adjust the learning content to their own needs and to find individual learning paths.

As it is hard to predict which learning style the user is favoring at the moment, we decided to make it possible to switch between the two navigation schemes at any time.

Alternatives. In an adaptive learning scenario, it is likely that alternative versions of the same learning content are available. The availability of these versions is based on the content (e.g. used media types) or contextual information. Although the application may initially make assumptions about which version would be most suitable for the learner, we came to the conclusion that it is important for the learner to freely switch between alternatives if he wishes to. The list of alternatives (if available) can be accessed by tapping on the content type icon of a card. For each alternative, the contextual information which is the condition for the learning unit as well as the content type is shown. Moreover, a thumbnail is shown to help the learner identify content that is most suitable for him.

Calibration. It can be necessary to pre-asses the learner to identify his prior knowledge. Unfortunately in a mobile learning scenario this is hard to achieve by questionnaire. On the one hand it is hard to create a generic questionnaire that fits all possible scenarios and on the other hand users usually aren't willing to endure a long question session just to consume rather short learning content.

To tackle this problem we decided to use the existing learning units as something similar to a questionnaire. After the on-boarding the learning units are listed to the user. As we are following the concept of micro-learning each learning unit will treat a small and manageable learning content. The user may look at a preview of the content of each unit and check off the units he already knows. This is by no means a perfect solution but rather provides as a basis for further research.

Another need for calibration arises from the fact, that contextual information may be subjective, e.g. ambient noise is perceived as disturbing by different users to varying degrees. The combination of gathered contextual information and user behavior already provides some hinds that can be used to automatically and manually adjust learning content to be selected in situations that are more suitable for learning.

In addition we decided to use the gathered contextual information for a given situation and occasionally ask if that situation is hindering concentration of the user (see Fig. 7). This provides a direct feedback by the user on which situations are ideal for learning and which are not.

Context change notifications. As was stated at the beginning of this section, it is important to inform the learner about newly gathered or updated contextual information. Above all, this feature aims at creating awareness for the contextual information that is gathered by the application. This, in return, enables the learner to identify privacy violations but also erroneous information.

In order not to distract the user from their actual learning task, the notifications were designed to be as non-obstructive as possible while maintaining a decent amount of information. As good as our intentions with these notifications are, we

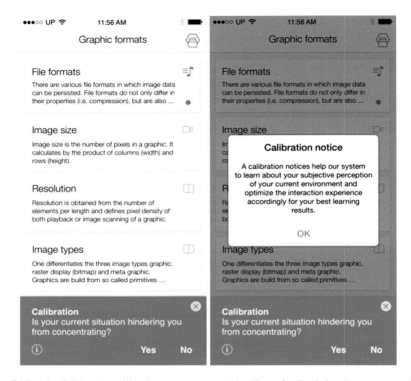

Fig. 7 On the *left* is the calibration prompt to gather direct feedback by the user on which situations are ideal for learning and on the *right* is a notification informing the user about the calibration

cannot assume that every user would like to be notified. Thus context change notifications will disappear automatically after a short while and will also present the user an option to disable notifications altogether.

What yet need to be evaluated is what kind of context information is suitable for these notifications and which notification frequency is acceptable for the user. It can be safely assumed that in most learning scenarios every second notification about changing ambient noise levels would soon be perceived as annoying (see Fig. 8).

Tracking events. As important as it is to make the user aware of changing contextual information, it is equally important to provide an easy to understand history of events and contextual information gathered in the past. Such a history function is represented by our *Trail* feature. The *Trail* chronologically keeps track of all past events and the contextual information that was gathered. In addition to an overview of the gathered data, the user can select single events or pieces of information and revise assumptions made by the context detection (e.g. delete information, mark information as false or change detection settings) (see Fig. 9).

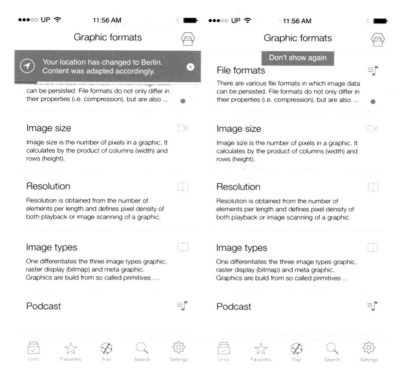

Fig. 8 An in-app notification informing the user that his current location has changed

5 Future Work

As the project is still in progress, some aspects will need some further work or refinement. The major open ends concerning context detection accuracy, cross-platform context detection, authoring user experience and mobile user experience will be addressed in this section.

5.1 Context Detection Accuracy

Some estimation about context detection accuracy that were made at the beginning of the project need to be reconsidered. That is no surprise since technology is in continuous change, and therefore makes ever more sources of contextual information available and may increase the accuracy of already detectable information, respectively.

For example, in the field of facial, emotions and gesture detection many new devices and technologies, like Oculus Rift, Myo, Leap Motion and Windows Hallo,

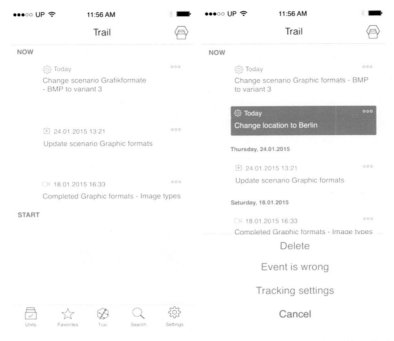

Fig. 9 The *Trail* feature chronologically keeps track of all past events and additionally lets the user revise assumptions made by the context detection

surfaced. These devices and technologies have made a lot of information available that had, until then, not been measurable at all. The trend towards wearable computing even promises many more sensors that will be available for use in the future.

Fortunately, our context detection was designed from the ground so that new contextual information can be added at any time. Nevertheless identifying new contextual information and incorporating software components to detect this information will be an ongoing task for developers that will be using the framework.

5.2 Cross-Platform Context Detection

Our context detection framework *contactJS* provides all means that are necessary to easily deploy a platform independent infrastructure to gather contextual information which can be used by mobile context-aware learning software to adapt its content to the preferences, knowledge and goals of its users.

Though contactJS in its present form is already a capable framework for the development of context-aware applications on mobile devices, further improvements are possible. This applies mainly to mechanisms that make the use of contactJS more convenient, such as the automatic detection and configuration of widgets and interpreters or the possibility of synonyms for certain contextual information. The latter will make it much easier to use widgets or interpreters that were created by developers who designed certain pieces of contextual information in a different way.

It is planned to release the context detection component (as well as the rest of the framework) as open source so that it is possible for the software development community to implement own features or context detection components.

5.3   Authoring User Experience

Developing a pleasant user experience for the authoring system is important because especially non tech-savvy users should be able to use the system. We are currently developing concepts and prototypes to evaluate a system that makes it possible to design mobile adaptive learning scenarios without the need to be firm with things like ontology reasoning or Boolean logic.

The main challenge is to map the abstract concepts of adaptivity and context detection, but also the relations between the various micro-learning units into an intuitive user interface. Furthermore design patterns need to be developed to support teaching staff to create effective micro learning content.

5.4   Mobile User Experience

The evaluation of the mobile user experience patterns that we developed is well underway. Initial feedback has already been positive, but also revealed some shortcomings. More tests are needed and are currently performed.

It has been found out that some of the design assumptions (e.g. using standard user interface components, or the support of irregular learning behavior) have proven to be right, it also became apparent that certain concepts (e.g. "Honeycomb" Navigation or onboarding) not completely satisfy the anticipated effects yet.

Regarding the core requirement of improving the transparency of context acquisition and adaptation, no definitive statement can be made. This requires more extensive testing with "real" students and "real" learning content. From the interviews with experts, however, it appears that context acquisition and adaptation are noticed neither positive nor negative. This can already lead to the conclusion that in order to achieve the goal of raising awareness of these mechanisms, further adjustments to the concepts will be needed. This could be made possible for example by an advanced primary onboarding with a clearer focus on adaptivity and context detection, as well as a secondary onboarding on learning scenario level to introduce in advanced methodology adaptation and scenario structure. Also the use of embedded help structures (e.g. coach marks) for interactive learning of application-specific navigation structures and to locate the different functions is to be classified as effective.

As our current patterns were developed on the basis of the iOS user interface, patterns need to be developed to transfer native GUI elements to other platforms (e.g. iOS tab bar to Android pop up menu).

6 Conclusion

Context-aware systems become more and more relevant for educational purposes. In particular, the trends towards mobile assistants with their broad spectrum of sensors and information retrieval techniques open broad possibilities for context acquisition and application on mobile devices.

The research visions of context aware systems for education or the utilization of context information for specific learning scenarios are impressive and innovative [9]. But, a transfer into actual, real world learning scenarios is often limited to dedicated educational settings.

This chapter was dedicated to the systematic realization of *mobile adaptive learning systems*. These are interactive systems that personalize and adjust e-learning content, pedagogical models, and interactions to the user's individual needs and preferences with a special focus on mobile end user devices for context acquisition. The perspectives of learners, teachers and developers on mobile adaptive learning systems have been discussed and evaluated—with a focus on the context data itself and user interfaces for learners and teachers.

A systematic approach for developing mobile adaptive learning applications has been introduced. In a first step, relevant contextual information has been identified by analyzing several educational scenarios from both, a technical point of view and a pedagogical perspective. In a second step, this context information has been classified regarding the current technical possibilities to acquire and utilize it an accurate manner. The findings of both steps led to a framework that supports even non-technophile authors or teachers in their creation of context-aware learning applications for mobile devices. Both, learner and teacher, benefit from highly intuitive interfaces that allow the utilization of such very technical concepts like context acquisition without requiring any deeper knowledge thereof.

However, the work on the framework continues, as current and future trends in e-learning offer potential for expansion and application. One of these trends, which represent a promising application for adaptive micro learning applications, is *Industry 4.0* (for example in the form of the resilient factory, intelligent maintenance management, networked production, etc. [42]). These scenarios typically provide an early and persistent need for training directly in the work process in order to limit costs and expenses to a manageable minimum. Adaptive mobile adaptive learning applications inherit the potential to satisfy these requirements.

Another promising current trend is the use of proactivity in software solutions. This means that an application is enabled to predict the behavior of the user and herby can offer information or answer requests before they have been explicitly stated. Prominent examples of this are so-called digital assistants (e.g. Google Now,[10] Cortana,[11] and Siri[12]). In the context of the framework, the combination of the collected context information and the actions of the learners could be used to predict future learning behavior, which in turn would allow for a more precise adaptation of the teaching content.

Finally, a recurring trend is competence-based assessment. Since competencies are commonly mapped in the form of ontologies, this may represent an elegant way to map prior knowledge as well as learnable skills into the form of contextual information. Current research deals with the development of cross-platform competence-based assessment [43].

Acknowledgments We would like to thank Helena Jank, Stefanie Lemcke, Martin Biermann and Julius Höfler for the invaluable work that they have contributed to the project. We would also like to thank the BMWi for the funding of the MOTIVATE project[13] and the Stiftung Industrieforschung for supporting Stefanie Lemcke's master thesis with a scholarship.

References

1. Hug, T.: Didactics of microlearning. Waxmann Verlag (2007)
2. Dey, A.: Understanding and using context. Pervasive Mobile Comput. **5**(1), 4–7 (2001)
3. Economides, A.: Adaptive context-aware pervasive and ubiquitous learning. Technol. Enhanced Learn. **1**(3), 169–192 (2009)
4. Ralston, A.: Interactive learning strategies for mobile learning (2013). http://www.unesco.org/new/en/unesco/themes/icts/m4ed/unesco-mobile-learning-week/speakers/anthony-ralston/
5. Shute, V.J., Psotka, J.: Intelligent tutoring systems: past, present, and future. Technical report, DTIC Document (1994)
6. Mitrovic, A., Koedinger, K.R., Martin, B.: A comparative analysis of cognitive tutoring and constraint-based modeling. In: User Modeling 2003, pp. 313–322. Springer, Berlin (2003)
7. Holt, P., Dubs, S., Jones, M., Greer, J.: The state of student modelling. In: Student Modelling: The Key to Individualized Knowledge-Based Instruction, pp. 3–35. Springer, Berlin (1994)
8. Lucke, U.: Pervasive learning. In: Pervasive Adaptation Research Agenda for Future and Emerging Technologies. Th. Sc. Community (2011)
9. Lucke, U., Specht, M.: Mobilität Adaptivität und Kontextbewusstsein im E-Learning. i-com **11**, 26–29 (2012)
10. Stoyanov, S., Kirchner, P.: Expert concept mapping method for defining the characteristics of adaptive e-learning: Alfanet project case. Edu. Tech. Res. Dev. **52**(2), 41–54 (2004)
11. Aguilera, N.E., Fernandez, G., Fitz-Gerald, G.: Addressing different cognitive levels for online learning. In: ASCILITE, pp. 39–46 (2002)
12. Pinkwart, N., Aleven, V., Ashley, K., Lynch, C.: Adaptive Rückmeldungen in intelligenten Tutorensystem largo. eleed 5(1) (2008)

---

[10]http://www.google.com/landing/now/.

[11]http://www.microsoft.com/en-us/mobile/campaign-cortana/.

[12]https://www.apple.com/ios/siri/.

[13]Funding reference: KF3155601MS3.

13. Aroyo, L., Dolog, P., Houben, G.J., Kravcik, M., Naeve, A., Nilsson, M., Wild, F.: Interoperability in personalized adaptive learning. J. Edu. Technol. Soc. **9**(2), 4–18 (2006)
14. Moebert, T., Jank, H., Zender, R., Lucke, U.: A generalized approach for context-aware adaption in mobile e-learning settings. In: 2014 IEEE 14th International Conference on Advanced Learning Technologies (ICALT), pp. 143–145. IEEE (2014)
15. Lucke, U., Rensing, C.: A survey on pervasive education. Pervasive Mobile Comput. **14**, 3–16 (2014)
16. Dillenbourg, P., Evans, M.: Interactive tabletops in education. Int. J. Comput. Support. Collab. Learn. **6**(4), 491–514 (2011)
17. Martyn, M.: Clickers in the classroom: an active learning approach. Educause Q. **30**(2), 71 (2007)
18. Uther, M., Zipitria, I., Uther, J., Singh, P.: Mobile adaptive call (mac): a case-study in developing a mobile learning application for speech/audio language training. In: IEEE International Workshop on Wireless and Mobile Technologies in Education, WMTE 2005, p. 5. IEEE (2005)
19. Giemza, A., Bollen, L., Seydel, P., Overhagen, A., Hoppe, H.U.: Lemonade: a flexible authoring tool for integrated mobile learning scenarios. In: 2010 6th IEEE International Conference on Wireless, Mobile and Ubiquitous Technologies in Education (WMUTE), pp. 73–80. IEEE (2010)
20. Oppermann, R., Specht, M.: Adaptive mobile museum guide for information and learning on demand. In: HCI (2), pp. 642–646 (1999)
21. Steimle, J., Brdiczka, O., Mühlhäuser, M.: Coscribe: integrating paper and digital documents for collaborative knowledge work. IEEE Trans. Learn. Technol. **2**(3), 174–188 (2009)
22. Zhao, X., Jin, Q., Okamoto, T.: Semantic retrieval: multiple response model for context-aware learning services. Int. J. Inf. Tech. Commun. Convergence **42**(3), 253–267 (2012)
23. Tortorella, R.A., Graf, S.: Personalized mobile learning via an adaptive engine. In: 2012 IEEE 12th International Conference on Advanced Learning Technologies (ICALT), pp. 670–671. IEEE (2012)
24. Weiser, M.: The computer for the 21st century. SIGMOBILE Mob. Comput. Commun. Rev. **3**(3), 3–11 (1999)
25. Schilit, B., Adams, N., Want, R.: Context-aware computing applications. In: Workshop on Mobile Computing Systems and Applications, pp. 85–90 (1994)
26. Vandewaetere, M., Vandercruysse, S., Clarebout, G.: Learners' perceptions and illusions of adaptivity in computer-based learning environments. Edu. Tech. Res. Dev. **60**(2), 307–324 (2011)
27. Schill, A., Springer, T.: Verteilte Systeme. Springer, Berlin (2007)
28. Dey, A., Newberger, A.: The context toolkit (2000). http://contexttoolkit.sourceforge.net/
29. Yamabe, T., Takagi, A., Nakajima, T.: Citron: a context information acquisition framework for personal devices. In: IEEE International Conference on Embedded and Real-Time Computing Systems and Applications (2005)
30. Bernados, A., Tarrío, P., Casar, J.: Casandra: a framework to provide context acquisition services and reasoning algorithms for ambient intelligence applications. In: International Conference on Parallel and Distributed Computing, Applications and Technologies, pp. 372–377 (2009)
31. Maia, M., Andre, F., Benedito, N., Romulo, G., Windson, V., Andrade, R.: Loccam-loosely coupled context acquisition middleware. In: 28th Annual ACM Symposium on Applied Computing, pp. 545–541 (2013)
32. Fahy, P., Siobhan, C.: Cass—middleware for mobile context-aware applications (2004). http://www.sigmobile.org/mobisys/2004/context_awareness/papers/cass12f.pdf
33. Hofer, T., Schwinger, W., Pichler, M., Leonhartsberger, G., Altmann, J., Retschitzegger, W.: Context-awareness on mobile devices—the hydrogen approach. In: 36th Annual Hawaii International Conference on System Sciences (2003)

34. Román, M., Hess, C., Cerqueira, R., Ranganathan, A., Campbell, R., Nahrstedt, K.: Gaia: A middleware infrastructure for active spaces. SIGMOBILE Mob. Comput. Commun. Rev. **6**(4), 65–67 (2002)
35. Hong, J.: The context fabric: an infrastructure for context-aware computing. In: CHI'02 Extended Abstracts on Human Factors in Computing Systems (2002)
36. Cheb, H., Finin, T., Joshi, A.: An intelligent broker for context aware systems. In: Adjunct Proceedings of Ubicomp, pp. 183–184 (2003)
37. Chen, G., Kotz, D.: Solar: an open platform for context-aware mobile applications. In: Proceedings of the First International Conference on Pervasive Computing, pp. 41–47 (2002)
38. Gu, T., Pung, H., Zhang, D.Q.: A middleware for building context-aware mobile services. In: IEEE 59th Vehicular Technology Conference, 2004. VTC 2004-Spring, vol. 5, pp. 2656–2660 (2004)
39. Bardram, J.: Tutorial for the java context awareness framework (jcaf), version 1.5 (2005). http://www.daimi.au.dk/~bardram/jcaf/jcaf.tutorial.v15.pdf
40. Leach, P., Mealling, M., Salz, R.: RFC 4122—A Universally Unique IDentifier (UUID) URN Namespace (2005). http://tools.ietf.org/
41. Silverman, W.L., Forum, L.: Learning and teaching styles in engineering education. Eng. Edu. **78**(June), 674–681 (2002)
42. Umsetzungsempfehlungen für das Zukunftsprojekt Industrie 4.0, Abschlussbericht des Ar-beitskreises Industrie 4.0, AGacatech – Deutsche Akademie der Technikwissenschaften e.V., April 2013, http://www.bmbf.de/pubRD/Umsetzungsempfehlungen_Industrie4_0.pdf (ARD Studie 2014) Online Medienutzung 2014; http://www.ard-zdf-onlinestudie.de/index.php?id=483)
43. Julian, D., Ulrike, L.: An infrastructure for cross-platform competence-based assessment. Accepted for: CHANGEE 2015. Facing the Challenges of Assessing 21st Century Skills in the Newly Emerging Educational Ecosystems. Toledo, 18 Sept 2015

A Revision of the Literature Concerned with Mobile, Ubiquitous, and Pervasive Learning: A Survey

Alejandro Peña-Ayala and Leonor Cárdenas

Abstract This chapter tailors a perspective of the work fulfilled in three learning research lines, which besides holding many common attributes also tend to converge to shape mobile, ubiquitous, and pervasive sceneries. Such a junction pursues to spread the traditional classroom and distance settings to open environments, as well as use the surrounding physical and digital objects as learning content that is available to learners at anytime, anywhere, and in any way. In sum, a complete learning environment is recreated to provide formal and informal learning to support academic studies, professional training, and lifelong learning. Thus, in this chapter a description of the mobile, ubiquitous, and pervasive learning (MUP-Learning) arena is presented through the selection of a sample of recent and transcendent works that offer from a conceptual contribution, such as models and frameworks, even empirical approaches oriented to specific domains of study. The sample of works is characterized according to a proposed pattern, as well as organized according to a suggested taxonomy. A profile to describe each work is also stated and a series of statistics are presented, as well as an analysis of the arena is provided to understand the potential and challenges related to the MUP-Learning field.

Keywords Mobile-leaning · Ubiquitous-learning · Pervasive-learning · Models · Frameworks

A. Peña-Ayala (✉)
WOLNM: Artificial Intelligence on Education Lab, 31 Julio 1859 No. 1099-B, Leyes Reforma, Iztapalapa, 09310 Mexico City, Mexico
e-mail: apenaa@ipn.mx

A. Peña-Ayala · L. Cárdenas
ESIME Zacatenco—Instituto Politécnico Nacional (IPN), Building Z-4, 2nd Floor, Lab 6, Miguel Othón de Mendizábal, S/N, 07320 Mexico, DF, Mexico
e-mail: adriposgrado@gmail.com

© Springer International Publishing Switzerland 2016
A. Peña-Ayala (ed.), *Mobile, Ubiquitous, and Pervasive Learning*,
Advances in Intelligent Systems and Computing 406,
DOI 10.1007/978-3-319-26518-6_3

Abbreviations

AR	Augmented reality
CSCL	Computer-supported collaborative learning
MUP-Learning	Mobile, ubiquitous, and pervasive learning
m-learning	Mobile learning
m-CSCL	Mobile computer-supported collaborative learning
p-learning	Pervasive learning
SRL	Self-regulated learning
SWOT	Strengths, weaknesses, opportunities, and threats
UNESCO	United Nations Educational, Scientific, and Cultural Organization
u-learning	Ubiquitous learning

1 Introduction

The continuous advancement of technology and trends toward mobile, ubiquitous, and pervasive computing has allowed its application in various domains, such as is education. Communications technology—which is supported by wireless devices that are broad connected and practically available everywhere—makes possible to take learning one step further. The exploitation of this sort of technology claims the conception on new pedagogical and learning paradigms to take advantage of its characteristics to recreate affordable teaching–learning environments. Thereby, the educational sceneries will benefit of the mobile, ubiquitous, and pervasive goodness.

The MUP-Learning field represents the junction point of three roads that hold their own personality as follows: the ability to deliver and get educational content at anyplace that distinguishes mobile learning (m-learning), the capacity to recreate context-aware sceneries that depicts ubiquitous learning (u-learning), and the opportunity to obtain information of the physical environment and its objects through the use of embedded devices at real time that correspond to pervasive learning (p-learning). Such benefits stage a nomadic, conscious, and living setting that facilitates the deployment of diverse learning paradigms, such as: blended, situated, immersive, etc.

In sum, MUP-Learning takes advantage of the strengths of the three afore-mentioned learning lines, as well as deals with their particular weaknesses, such as: the low embeddedness of m-learning, the lack of a widely accepted learning theory for u-learning, and the location dependency that accuses p-learning. Thus, in order to overcome those deficiencies, MUP-Learning aims at designing effective assessment, apply suitable pedagogic paradigms, and build new learning applications that include diverse functionalities related to augmented reality (AR), computer-supported collaborative learning (CSCL), social networking, data mining, self-regulating learning (SRL), user modeling, learning analytics, educational robotics, metacognition, etc.

Therefore, with the purpose to sketch a view of the current MUP-Learning state of the art, this chapter is organized according to the following structure: In Sect. 2 a conceptual frame of the field is given by the exposition of the MUP-Learning background and the statement of essential concepts oriented to lay the foundations of the field, as well as a collection of previous reviews related to the three fields. Whereas, in Sect. 3 the logistic behind the review is unveiled through: the statement of the research questions, the identification of the criteria used to seek sources, and the illustration of the frameworkk that guides the survey, as well as a twofold proposal, one as a taxonomy to classify the works and a pattern to characterize them.

Regarding the results, these are detailed in Sect. 4, as a sample of 105 works published from 2010 up to the first quarter of 2015; where, such a collection is organized according to the categories and subcategories that compose the taxonomy. While in Sect. 5, the discussion of the results is given by the exposition of diverse statistics and the identification of some findings, as well as the analysis of the strengths, weakness, opportunities, and threats (SWOT). Finally, the conclusions are outlined in Sect. 6 as the future work to be carried out and the definition of some research trends, including the provision of the responses to the research question.

2 A Glance at Mobile, Ubiquitous, and Pervasive Learning

This section pursues the goal to contextualize the MUP-Learning review. Thus, three topics are stated to base the exposition of a landscape for MUP-Learning. First, a brief chronicle of previous learning systems is outlined in order to acknowledge the background of the MUP-Learning. Afterward, the definition of some essential MUP-Learning concepts is given to put the reader in context. Finally, a series of previous reviews related to some of the MUP-Learning lines is summarized.

2.1 Learning Systems in a Nutshell

With the purpose of introducing the context of the review, a summary of diverse educational systems is outlined in this subsection to acknowledge the background that provides diverse foundations for MUP-Learning systems.

According to McDonald et al. [1], the pioneer efforts at automating instruction date back to the early nineteenth century. Later on, in the 1920s, Sidney Pressey was interested in improving the learning process and teacher labor, so he promoted the programmed instruction method and invented the "Automatic Teacher Machine" [2–4].

In the 1960s, Harvey Long used an International Business Machines 1440 mainframe to set up a nationwide online learning system to train engineers throughout the country. In that time, users interacted through a telecommunications system by teletypewriters and were tracked on their progress through the programs [5]. As a result, the computer-aided instruction emerged to develop teaching systems, which could adapt to the needs of individual students [6]. Subsequently, in the beginning of the 1970s, intelligent tutoring systems arose to behave like problem-solving monitors, coaches, laboratory instruments, and consultants [7].

Later on, during the 1980s personal computers, as well as local access networks and wide access networks appeared in academic, professional, and home settings. In consequence, new educational systems were born, such as: educational hypermedia systems [8], CSCL [9], and learning management systems [10]. Afterward, as a result of the Internet invasion, educational systems migrate to a wide sort of web-based paradigms, where some of them correspond to: web-based education [11], intelligent learning environments [12], m-learning [13], social networking [14], p-learning systems [15], educational data mining [16], and u-learning scenarios [17].

2.2 Definition of the MUP-Learning Field

MUP-Learning is a term coined to highlight an enhanced learning environment that is the result of the confluence of three research lines, whose technological and functional riches is higher than the ones that each of them could individually reach. Thus, in order to sketch a statement for this new term, it is desirable to take into account illustrative definitions provided for the three converging lines (e.g., m-learning, u-learning, and p-learning) as follows.

M-learning. In relation to m-learning, its baseline is the mobile computing, also called *nomadic computing*, to mean the use of portable computing devices in conjunction with wireless communications technologies to enable users to synchronous and asynchronous access and transfer data, interact with applications, and collaborate with peers. According to Taniar, mobile computing sits in the joint of motion and computing, where users have the ability to move long distances and electronically perform computing tasks, both in a few time have changed and opened the entire world [18].

Concerning to m-learning, UNESCO expresses that: it involves the use of mobile technology, either alone or in combination with other information and communication technology, to enable learning anytime and anywhere [19]. In addition, Lucke and Rensing state: m-learning focuses on the mobility of a learner, who uses information and communication technologies and emphasizes the learner's movements [20].

U-learning. As for u-learning, a key underlying element is the ubiquitous computing, whose concept was envisioned by Mark Weiser in 1991 to express: the next-generation computer technologies that weave themselves into the fabric of everyday life until they are indistinguishable from it [21]. In other words, people should be able to work with computing devices without having to acquire the technological skills to use them. In addition, Peng et al. express: "The term ubiquitous computing means on-demand computing power with which users can access computing technologies whenever and wherever they are needed" [22]. In short, the aim of ubiquitous computing is to create a new relationship between people and computers in which the computers are kept out of the way of users as they go about their lives.

In so far as u-learning is defined by Hwang et al. that "any learning environment allowing learners to access content at anywhere and anytime, no matter whether wireless communication or mobile devices are employed" [23]. Authors also express that context-aware u-learning refers to learning with mobile devices, wireless communications, and sensor technologies that together enable the environment to be conscious of the user, as well as users are immersed in an authentic learning experience.

P-learning. Concerning p-learning, this line is supported by pervasive computing, which in accord with Alsiyami: pervasive computing is a model of human–computer interaction where information processing integrates into everyday activities. It is a rapidly developing area and has many potential applications from domestic ubiquitous computing to environmental monitoring and intelligent spaces [24].

In words of Sherimon and Reshmy, p-learning is learning enhanced with intelligent environments and context awareness. They express that the information about the learner's context is obtained from the learning environment which is embedded with sensors, tags, and so on [25]. What is more, Lucke and Rensing point out that there is a broad consensus that context, context awareness, and adaptivity can be seen as the core concepts of p-learning [20].

MUP-learning. Regarding MUP-learning, this term is grounded on the aforementioned concepts. Thus, MUP-Learning tends to recreate a kind of learning experience, where learners freely move along indoor and outdoor settings using mobile devices, which deliver ubiquitous educational contents available at anytime, anywhere, and anyway in an enhanced and rich learning scenery supported by embedded and context-aware technologies with the purpose to perform educational tasks.

2.3 Previous Reviews

MUP-Learning has been studied and surveyed from diverse perspectives as the following sample of reviews shows. The first review is conducted by Song [26] who investigates methods applied in m-CSCL which focus on studying, learning, and collaboration using mobile devices. The author identified whether these methods have examined m-CSCL effectively, when they are administered, and what methodological issues exist in m-CSCL studies.

Regarding the second review, it is introduced by Gilman et al. [27], who as a result of defining the u-learning state of the art, identify four user roles: learner, instructor, developer, and researcher. Authors characterize the way learner plays a specific role. Moreover, they claim supporting different needs to facilitate the achievement of the user roles by adding the meta-level functionality to u-learning environments.

In relation to the third review, Pachler et al. present an overview and critique of theoretical and conceptual frames currently used to explain and analyze learning with mobile devices [28]. They discuss the most pertinent theoretical perspectives and explanatory conceptual frames with a view to exploring their respective characteristics, strengths, and weaknesses.

Lastly, a fourth review is conducted by Lucke and Rensing [20], who present an overview of the existing work in the application of mobile, ubiquitous, pervasive, contextualized and seamless technologies for education. The authors address pedagogical and technological issues. Inclusive, they identify areas within pervasive education that are currently disregarded or deemed.

3   The Survey Method

In this section, the logistic followed to carry out the review of the MUP-Learning field is introduced, as well as the material exploited to shape the survey. So, the kick off corresponds to the research questions that inspire the work. Later, the traits of the source to collect the sample are uncovered. Afterward, a sketch of the workflow applied to tailor the survey is illustrated. A couple of contributions close the section to introduce a taxonomy to classify MUP-Learning works according to their nature and purpose as well as a pattern to characterize them.

3.1   Research Questions

With the purpose to picture a landscape of the MUP-Learning, the following series of research questions is made to settle on the work to be fulfilled:

1. What are the basics of the field?
2. How to classify the works to meet different perspectives?
3. How to characterize the works?
4. What is the work currently achieved?
5. Which are the most frequent topics addressed?
6. What are the highlights of the field?

3.2   Sampling: Sources and Works

In order to outline a fresh and relevant perspective of the MUP-Learning labor, a set of criteria are defined to make up the sample. Based on such constrains, several hundreds of works were gathered, but only 105 met the criteria to compose the sample. The alluded traits that the works presented in this review satisfy are the following:

- Recent works, those published since 2010 up to the first quarter of 2015.
- Preferably, works published in journals indexed by Thomson Reuters Journal Citation Reports® and Social Science Citation Index, and specialized books.
- Exclusively empirical and conceptual works, as well as domain oriented.

3.3   Framework of the Present Work

In relation to the method applied to shape the survey of MUP-Learning works, it is sketched as the framework illustrated in Fig. 1. Where, 12 tasks, the aforementioned criteria, and a couple of databases are organized as the next sequence:

1. Definition of the selection criteria.
2. Gathering of related works.
3. Selection of works.
4. Definition of a taxonomy to classify the chosen works.
5. Classification and organization of the sample works.
6. Definition of a pattern to characterize the selected works.
7. Description of the works through a summary that shows their nature and scope.
8. Statistical process to highlight the main attributes of the works.
9. Data mining discovery.
10. Analysis and interpretation of the outcomes.
11. Discovery of findings.
12. Edition of the review.

3.4   A Taxonomy to Classify the Related Works

Due to the nature of the surveyed works, it is evident authors consider diverse factors to orient their studies such as: purpose, settings, context, technology, or experiment. So, a broader perspective to describe the related works is necessary to classify them.

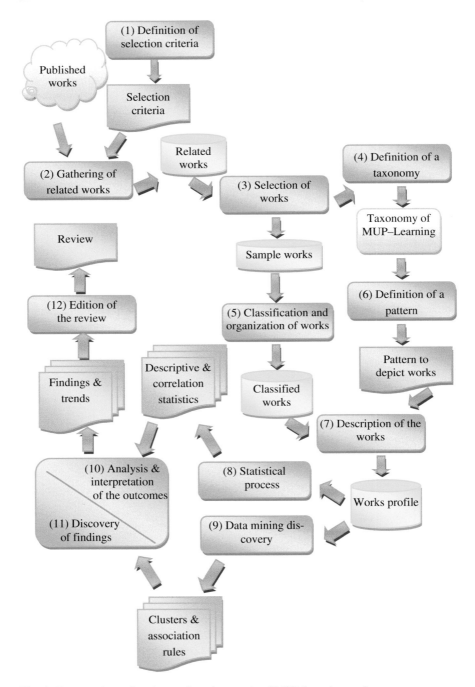

Fig. 1 Framework employed to analyze the sample of MUP-Learning works

Table 1 Taxonomy proposed to classify MUP-Learning works

Category	Subcategory	Topics
Conceptual and empirical (53)	Models (13)	
	Frameworks (17)	
	Tools (6)	
	Approaches (17)	
Domain studies (52)	Sciences (17)	
	Language learning (7)	
	Attitudes (13)	
	Education (6)	
	Diverse fields (9)	Messaging (2)
		Arts and architecture (4)
		Workplace (3)

In this context, the taxonomy, sketched in Table 1, is proposed to organize the sample. It is a three-tier hierarchy, which embraces two *categories* to gather common views. They are also split into *subcategories* to define a given view. Inclusive, one subcategory contains specialized topics to precise the nature of the work.

Besides the hierarchy that shapes the taxonomy structure and identifies the three levels of categories, the count of related works that correspond to each of them appears inside of parentheses. The sum of the count for the two main categories is 105, whereas their respective count is the result of the accumulation of its descending subcategories' counts. In this way, it is evident the symmetry between the two branches.

3.5 A Pattern to Characterize the Sample Works

According with the aforementioned taxonomy and the frequencies estimated for each topic, the following seven variables are defined to organize the pattern, and they are instantiated in the Table A.1 of the Appendix:

1. Citation: A number that identifies the work.
2. Year: Value between 2010 up to 2015.
3. Source: Journal, book, proceedings: Acronym of the source.
4. Country: Nation where is the affiliation of the author.
5. Academic level: It is a standardized term to facilitate clustering.
6. Learning setting: Environment where the work is carried out.
7. Domain: It is a standardized term to facilitate clustering.

4 A Landscape of Mobile, Ubiquitous, and Pervasive Learning

With the aim at tailoring a view of the recent labor accomplished in the MUP-Learning arena, promote its outcomes, and encourage future research, this section offers a collection of relevant works. Such a series is introduced according to the taxonomy to classify MUP-Learning works, exposed in Sect. 3.4, by means a brief description of the category and a summary of representative works.

4.1 Conceptual and Empirical Works

The work achieved in the MUP-Learning is heterogeneous because some works offer a conceptual proposal to model and guide the MUP-Learning labor. Whereas, several works sketch diverse approaches to support the development and implementation of diverse MUP-Learning applications. Inclusive, a sample reports empirical experiences as a result of the testing and deploying of an approach to tackle a specific subject.

Models. These kinds of works propose or apply a sort of model to characterize learning development, as well as sceneries as part of the approach they report. Thus, some interesting contributions are reported as follows:

- Huang and Chiu [29]: analyzes the effectiveness of a meaningful learning-based evaluation model for context-aware m-learning. They take into account the meaningful learning theory using an analytic hierarchy process. In order to verify the effectiveness of the model, three different context-aware m-learning learning activities are tested, where diverse mobile devices are evaluated, such as: learners collaborate to study plants in an m-learning context, students learn about plants on a botany course, and learners analyze trees on campus as the context-aware m-learning field.
- Lee [30]: believes the use of technology for learning would necessarily require learners to exercise regulation over their course of actions, especially when technology is an integral part of the education landscape. Thus, author pursues to tailor a model to correlate regulation of cognition and the intention to use technology as a visualization tool.
- Sha et al. [31]: conceptually and empirically explore how theories and methodologies of SRL help to analyze and understand the processes of m-learning. So, authors design an analytical SRL model of m-learning to link three key aspects: students' self-reports of psychological processes, patterns of online learning behavior in the mobile learning environment (MLE), and learning achievement.
- Mills et al. [32]: are interested in informal learning from three views, information seeking, information sharing, and going mobile. In consequence, they

use the guided inquiry spaces model [33] as a reference to bridging the student's informal learning world and the curriculum-based teacher's world.

- Liao et al. [34]: explore the antecedents of collaborative learning performance over social networking sites in a u-learning context. Researchers modify the technology acceptance model [35] and augment it with external factors, including collective efficacy and personal innovativeness in information technology. This new version of model is then used to examine the influential factors in students' use of social networks to learn, and also to evaluate their learning attitudes and usage effects.

- Wagner et al. [36]: proposes a model to register entities' actions in trails and infer profile information from these trails, using semantic interoperability. Also, it allows applications to share information and infer a unified profile. Such a model was used as part of an application in a u-learning scenario, where the students' profile was dynamically updated, allowing them to better adapt to the environment.

- Chen et al. [37]: uses a theoretical instructional pervasive game model to construct a cultural-based pervasive game in a p-learning environment. It pursues to support individual and collaborative learning methods, enhance learning effectiveness, encourage attitudes toward mobile devices, and increase user's satisfaction.

- Arnone et al. [38]: focuses on the learners' curiosity, interest, and engagement at interacting with p-learning scenarios. Thus, authors sketch a theoretical model for curiosity, interest, and engagement in new media technology p-learning environments taking into consideration personal, situational, and contextual factors as influencing variables.

- Çalik et al. [39]: evaluate the effects of environmental chemistry' elective course via technology-embedded scientific inquiry model [40] on senior science student teachers' conceptions of concepts, issues, attitudes, and technological pedagogical content knowledge levels. Such a model has two traits: (a) interactions among three hallmarks of scientific inquiry (e.g., scientific conceptualization, investigation, and communication); (b) the use of educational technologies to conceptualize, investigate, and communicate in the process of inquiring.

- Reychav and Wu [41]: based on the uses and gratification theory [42] conceive a research model to understand how learning outcomes are influenced by user information needs for training, innovativeness, new and cool perceptions of using mobile technologies, user preferences, and user perceived enjoyment. A structural equation model [43] analysis was run to test the research model during the exploration of mobile tablet training for road safety.

- Boyce et al. [44]: helps students to experience the natural world, to enhance their understanding of science, and to pique their interest in science. So, they apply the attention participation model [45] to place students within an informal learning setting that split their attention and participation across three planes: (1) social (i.e., interactions among students and interactions between a student and naturalist); (2) technology (i.e., interactions between a student and an iPad app); (3) nature (i.e., interactions between a student and the natural environment).

- Al-Hmouz et al. [46]: propose a model of an adaptive neuro-fuzzy inference system for delivering adapted learning content to mobile learners to satisfy individual learner traits by considering a learner's learning style. They use their own enhanced learner model [47] that depicts four status: learner, situation, knowledge and shared properties, and educational activity. In order to reach m-learning adaptation, seven adaptation layers are considered: context acquisition, information classification, learner model, information extraction, learner profile representation, reasoning, and interface.

- Kali et al. [48]: offer an instructional model that conducts collaborative inquiry in situ and streamlines learning between classroom, museum, and home to support undergraduate-level art history students in developing the skills required for analyzing artwork. Such a model was studied in three aspects: (1) its potential to enable instructors to implement the cognitive apprenticeship instructional approach; (2) its contribution to the students' development of independence and self-efficacy in analyzing artwork; (3) the contribution of technology to streamlining learning between settings.

Frameworks. With the purpose to guide the development of a whole MUP-Learning, or at least a given functionality, researchers propose methods, procedures, and techniques that orient the work to be done. In consequence, a collection of proposals as well as their application of some of them is introduced as follows.

First, practical guidelines for designing instructional messages for m-learning are outlined in [49], where authors take into account the influence of learning and cognitive theories, human–computer interaction principles, devices, and methodologies. Moreover, a conceptual framework for sustainable m-learning in schools is pointed out in [50]. It re-contextualizes the human factors that are embedded in Cisler's framework [51] upfront to enable the theorizing of sustainable m-learning programs to explore how the stakeholders interact with each other and with the technology.

As for [52], authors propose a principle-based pedagogical design framework for developing constructivist learning in a seamless learning environment, where personal mobile devices and social learning networks are used. The aim is to tailor a teacher developmental model for constructivist learning and teaching in digital classrooms. In another vein, [53] applies the integrative learning design framework for design-based research [54] (i.e., it contains the next stages: informed exploration, enactment designing, evaluation the local as well as the broader impact) in order to embedding computer supported collaborative learning (CSCL) by means of handhelds.

In regards [55], Green et al. report the use of the m-learning pedagogical framework tailored by Kearney et al. [56] to design a mobile app selection for science rubric. Their aim is to analyze the students' processes and products of the artifact creation in out-of-school settings.

In the meantime, [57] implements m-learning curricula in schools by means of a task-interaction and systemic framework to support educational decision-making in m-learning based on the relationships between the interactions occurring in a learning activity and the tasks which are pedagogically relevant for the activity.

Concerning to Yau et al. [58], state a mobile context-aware framework for managing learning schedules in order to suggest content to students based on the values of the proposed contexts related to learning styles, knowledge level, concentration level, and frequency of interruption at the point of usage.

Whilst, [59] deploys a framework of participatory simulation for m-learning using goals and scaffolding to enhance students' experiential learning. It adopts the Kolb's experiential learning model [60], which consists of five sequential but cyclic steps: initial, concrete experience, observation and reflection, abstract conceptualization, and testing in new situations.

In relation to [61], it develops a cyclic process to support seamless learning of language by the production of linguistic artifacts, which embraces four activities: (1) in-class/on-campus contextual idiom learning; (2) out-of-class, contextual, independent sentence making; (3) online collaborative learning; (4) in-class consolidation. While [62] follows a design-based research approach to address complex issues in classroom contexts in collaboration with practitioners, and integrates design principles with technological affordances to create solutions to meet teaching and learning needs.

On the other hand, [63] proposes a principle-based pedagogical design framework for inquiry-based learning in the seamless learning environment. The aim is to suggest contents to students based on the values of the proposed contexts including learning styles, knowledge level, concentration level, and frequency of interruption. In addition, [64] presents a systematical approach for developing a framework for context-aware adaptive learning by analyzing several different educational scenarios from a pedagogical point of view and the way they can be adapted to the learner.

As regards with [65], authors develop a mobile integrated and individualized course by their own process to design adaptive educational systems, which follows four stages: (1) inputs: assessment points, score predictions, viewing behavior… (2) modeling: point tracking, collaborative filtering, correlation… (3) path selection: static, step by step, sequencing; (4) path generation: manual, hybrid, automatic. What is more, [66] describes a framework to address the design of smart learning environments to support both online and real-world learning activities from the perspective of context-aware u-learning, which encompasses the following modules: learning status detecting, learning performance evaluation, adaptive learning task, adaptive learning content, personal learning support, learner profiles, and an inference engine.

Moreover, [67] shapes a framework for analyzing learners' perceptions and responses that resulted from physical environment changes by investigating the dynamic factors of classroom and five elements of learner experience (e.g., value, usability, adaptability, desirability, comfortability). Inclusive, [68] designs a meaningful learning-based evaluation method for u-learning along both macro and micro aspects, and in an effort to make u-learning more sustainable by a series of suggestions for instructors and designers interested in promoting the quality of u-learning. In addition, [69] proposes a method, called "personalized learning content adaptation mechanism", to meet diverse m-learning users, which applies clustering and decision tree algorithms to manage historical learners' requests that are interpreted to deliver personalized learning content.

Tools. Other contributions made by MUP-Learning researchers corresponds to the development of tools specialized in a specific task. As a sample of such works, Table 2 illustrates some of their characteristics.

Approaches. With the purpose to illustrate the empirical labor that is being carried out in the MUP-Learning arena, a series of approaches are described as follows to highlight their nature and scope, as well as their outcomes.

- "Picaa" [77]: is an approach to assist students with special needs that uses iPad and iPod to cover the learning processes of preparation, use, and evaluation by means of four kinds of educational activities: exploration, association, puzzle, and sorting.

Table 2 MUP-learning tools to support the development of specific functions

Tool	Name	Nature
[70]	m-learning tool designed for self-assessment	Tool designed to reinforce students' knowledge by self-assessment. It was tested by measuring improvements in student achievements and applying an attitudinal survey to estimate student attitudes towards this new tool
[71]	Mobile digital armillary sphere	It is designed based on kinesthetic learning style theory, interviews regarding teachers' experiences applying traditional astronomy teaching methods, and the use of AR to support astronomical observation instruction
[72]	Lifelong learning hub	This is a mobile seamless tool for SRL that aims to cover lifelong learners' learning process: feedback, learning activities across locations, linking between learning activities and everyday life, incompatibility between near field communication tags and readers
[73]	dmTEA	The tool deploys the inventory of the autist spectrum [74] in order to allow both behavior evaluation and modelling of students on the autism spectrum by means of 12 dimensions which are the main disorders that define it
[75]	Interactive dashboards	The tool should be created to display information and assist teachers in real time, as well as for research purposes and decision-making with the aim at maximizing teacher's time which become diffused in u-learning scenarios
[76]	Collaborative mobile learning	It is based on computer-mediated communication tools to aid collaborative problem-solving by enabling the language of science to be modeled for knowledge-building

- Helping autism-diagnosed teenagers navigating and developing socially [78]: is a project to develop social and life skills in children with autism spectrum disorders that applies the persuasive technology design [79] to offer mobile cognitive support and measure how engaged children are with technology.
- Development of social and life skills in children with autistic spectrum disorders [80]: represent a series of key factors to be taken into account to mediate the use of mobile technology and determine the level of influence it has on practice.
- Automatic text summarization approach [81]: assesses learning outcomes associated with reading text summaries in mobile learning contexts, where full text learning contents or various summaries are displayed to learners.
- Smart classroom system that integrates near field communication [82]: automates attendance management, locates students, and offers real-time student feedback, as well as evaluates its effect on students' attitude toward science education.
- Responsive and adaptive web for open and u-learning [83]: guides the content authoring in language learning to adaptively respond to user's media and enable interfaces to dynamically suit the user device without the use of specialized modules.
- Adaptive u-learning system [84]: provides context awareness and courseware recommendation functions to enhance lifelong learners' effectiveness to complete their learning task more quickly and more accurately to achieve personal goals.
- U-Base [85]: is a ubiquitous decision support system that collects user transaction data to construct general Bayesian networks and predict user's future contexts, as well as provide context-sensitive recommendation to users.
- Context-aware and personalized event recommendation [86]: is an approach built as a multi-agent architecture that uses an ontology and a spreading algorithm to respectively define the domain knowledge model and learn user interests patterns.
- Running Othello [87]: is a distributed pervasive exergame that recreates a pervasive world by combining a board game with sensor-enhanced physical activity that is based on cognition and learning, exercise, gaming, and sociality principles.
- Context-aware mobile educational game [88]: is based on a multi-agent architecture to generate a series of learning activities for users doing on-the-job training and make users interact with learning objects in their working environment.
- Learning log navigator [89]: is a function of the system for capturing and reusing of learning logs that allows learners to record and share their daily learning experiences as ubiquitous learning logs to tailor a task-based language-learning setting.
- Natural language interaction [90]: is an approach that applies semantic analysis to enhance question and answer processing, automated question answering, and automatic text summarization involved in u-learning systems to benefit learners.

- Mobile traffic violation reporting system [91]: integrates mobile communication technologies and a global positioning system to improve learners' traffic violation reflection level with the aim to increase drivers' lay consciousness.
- Personalized guide recommendation system [92]: pursues to mitigate information overload in museum learning by discovering and proposing recommendation rules both from collective visiting behavior and individual visiting behavior.
- "iSpot" [93]: is a project oriented to encourage people to explore, study, enjoy, and protect their local environment by two modalities, iSpot local and iSpot mobile, which respectively correspond to location-based activity and m-learning.
- Blended mobile museum learning environment [94]: this approach examines the learning process, learning performances, and behavioral patterns of visitors when they walk through the museum accompanied by a learning website under a typical style, using paper-based learning sheets, and interacting with m-learning systems.

4.2 Domain Studies

MUP-Learning concerns with the provision of domain content specialized in a given discipline, as well as the development of certain cognitive skills and attitudes in formal and informal settings. Thus, the remaining collection of works is organized according to the taxonomy stated in Sect. 3.4. Thus, this subsection is split into five parts, where some of them embrace several subcategories.

Sciences. One of the main concerns of MUP-Learning corresponds to the teaching of sciences, which usually demand a special learners' effort to be well understood. So, innovative paradigms should be implemented in MUP-Learning scenarios to facilitate the acquisition of domain knowledge as the following works demonstrate.

Charitonos et al. [95] investigate the use of social and mobile technologies in school field trips as a mean of enhancing the museum visitor experience. They explain the role of such resources in fostering the social interactions around museum artifacts and ultimately the process of shared construction of meaning making. Other context-situated work corresponds to [96], where Ruchter et al. study the impact of a mobile guide system on different parameters of environmental literacy in comparison to traditional instruments of environmental education (e.g., brochure, human guide).

As inquiring, Jones et al. [97] aim at understanding more about learner control and how technology promotes learners' inquiries. Thus, they examine how e-Learning supports science inquiry learning by adolescents in a semiformal context, and the way adults use their own mobile technologies to learn about landscape in informal settings. Whereas, [98] introduces the project "Bring Your Own

Device", where it is investigated what advancement of content knowledge students made in their science inquiry in a seamless learning environment supported by their own mobile device.

What is more, [99] investigates and compares students' collaborative inquiry learning behaviors and their behavior patterns in an AR simulation system and a traditional two dimensions simulation system, where their inquiry and discussion processes are analyzed by content analysis and lag sequential analysis. In addition, [100] examines small groups of students who are using mobile devices in authentic educational settings, within a natural science inquiry-based learning activity outdoors. To this end, authors characterize learning activities mediated by mobile technologies through the second-generation cultural-historical activity theory [101].

Whilst in [102], an attention-to-affect model with a self-report measure are used to determine the antecedent factor, Internet cognitive failure, related to learning interest based on students' continuance intentions to observation, explanation, and prediction. Also, [103] deploys wireless sensor networks for tutoring courses offered in laboratory, where the students' experimental skills and learning inclinations are evaluated with a questionnaire that assesses skills of taking initiatives in: experimental inquiries, setting experimental hypothesis, experimental procedures, collaborative interactions exchanged, and the questioning and introspections reflected upon the findings.

As for games, Furió et al. reinforce children's knowledge about water cycle [104]. The game included touch screen, accelerometer and combined AR mini-games with non-AR mini-games for better gameplay immersion. What is more, [105] studies the effect of using a mobile literacy game, called "GraphoGameTM", to improve literacy levels. The assessed of the achieved learning is made through cognitive tests that measure emergent literacy skills, decoding competence, vocabulary, and arithmetic.

As to natural sciences, [106] explores users interactions in a learning environment designed to support science learning outdoors at an arboretum and nature center using mobile devices, called the "Tree Investigators". Thus, authors coded video records and artifacts created by children and parents to understand how both social and technological supports influenced observations, explanations, and knowledge about trees.

Furthermore, [107] measures student attitudes and knowledge in technology-rich Biology classrooms with the aim at discovering any differences based on gender after a course. As a result of applying the "students' attitudes toward and knowledge of technology questionnaire", it was found out statistically significant gender differences in all the scales of the questionnaire in favor of males.

Whereas [108] examines the effects of prior knowledge on learning from different compositions of representations in m-learning environments on plant leaf morphology. So, two experiments compare (1) the learning effects of an m-learning environment presenting text and photos of plants on a tablet PC, either in combination with or without real plants in the physical environment; and (2) the differential effects of prior knowledge on learning with the combination of texts, photos and real plants to a combination in which the photos were replaced by schematic hand drawings. Results show that subjects of the second trial performed better on a comprehension and an application test.

In addition, [109] explores how connected classroom technologies facilitate formative assessment. Thus, authors analyze the experiences of science teachers in their first year of implementing the "TI-NavigatorTM" system. They conclude that such technologies support the implementation of a variety of instructional tasks that generate evidence of student learning for the teacher on which to base subsequent instructional decisions. By the side of [110], the research evaluates the effectiveness of podcasts delivered as-needed basis on iPhones or iPod in a chemistry laboratory setting. Such podcasts with audio and visual tracks cover key laboratory techniques and concepts that aid in experimental design. The study shows that student laboratory teams were able to gather laboratory information more effectively when it was presented in an on-demand podcast format than in a pre-laboratory lecture format.

Other science work is reported in [111], where it is thought that three dimensions-based simulations can be used to reverse learners' misconceptions of abstract topics. Therefore, authors apply the idea to facilitate the conceptualization of astronomical scales in order learners to advance their understanding of the solar system. They found out that even brief exposures to instruction based on pinch-to-zoom simulations of the solar system advanced students' understanding in areas where traditional instruction is notoriously ineffective. Lastly, [112] compares online multimedia learning with mobile devices and desktop computers. So, authors develops a trail where learners receive an online multimedia lesson on how a solar cell works, some of them at lab using computers, and others in a courtyard working with iPad's. As a result, the mobile users group outperformed the desktop group on a transfer test.

Language Learning. A popular demand represents the use of MUP-Learning technology to teach language for users of all ages, in outdoors settings, and as a topic for lifelong learning as well as a viable subject for blended learning scenarios. A sample of the work achieved in language learning, Table 3 offers a profile of seven applications, where the first two correspond to support primary school students, the third and fourth are devoted to middle, the fifth to high school learners, and the sixth and seventh to ungraduated students and informal learning respectively.

Attitudes. Inner learners' world plays an essential role in the success of MUP-Learning achievements, where feelings, motivations, and engagement represent just a sample of the mental factors that determine learners' behavior and outcomes. Thus, in this section a relation of 13 studies is outlined to unveil key psychological factors.

- *Perceptions* are investigated in a study concerned with the preservice teachers' perceptions about using m-phones and laptops in education as m-learning tools [120]. The attitudes towards using laptops were more positive than mobile phones. As for the limitations, the situation was balanced for using laptops and m-phones.

Table 3 A collection of MUP-Learning works oriented to language learning

Tool	Scope	Nature
[113]	English comprehension and vocabulary acquisition	Studies the effects of non-caption, full-caption, and target-word display modes on English learning. It is found that the learning achievement of the English target-word group was as good as that of the full-caption group in terms of vocabulary acquisition
[114]	Chinese character learning	Shows the effects of spontaneous learners grouping to develop orthographic awareness at learning Chinese characters by means of a mobile game. The game process data and the transcriptions of focus group interviews were analyzed to identify the dynamics of student collaboration and competition during the games
[115]	English learning	Reveals a study of English language learners, teachers and students who speak diverse languages in addition to English, work with iPod touches 24/7. The results unveil diverse academic challenges in language acquisition as well as the benefits of such devices as a teaching and learning tool for this kind of population
[116]	English vocabulary learning	Analyzes the effects of small, medium and large screen sizes, as well as text only and text with pictorial annotation of multimedia instruction on vocabulary learning. Results show that the large screen helped the students to learn more effectively than the small screen
[117]	English learning	Investigates the roles of mobile technology playfulness, users' resistance to change, and self-management of learning in mobile English learning outcomes. The questions of interests were related to the usage of electronic dictionaries and their effects on m-learning outcomes
[118]	Second and foreign language learning	Examines the current state of learners' attitudes toward mobile technology use (i.e., if age, gender or cultural factors affect these attitudes). Findings show the respondents' attitudes toward m-learning are positive with individualization being the most followed by collaboration
[119]	Dutch language learning	Exhibits the outcomes of a study of mobile media delivery for language learning, comparing two context filters and four selection methods for language content. The groups were compared on knowledge gain, and the outcomes indicated that the results differed significantly

- *Perceptions* are examined in the light of teachers' age; particularly, how the type of mobile phone is related with them, their support for the use of mobile phones in the classroom, their perceptions of the benefits of mobile features for school work, and their perceptions of instructional barriers [121]. The results unveil that the age of the teacher matters, due teachers over 50 years old were less likely to own smartphones, less supportive on all items, less enthusiastic of the features, and present barriers that make them more problematic users than younger teachers.
- *Perceptions* are pondered from surveys applied to faculty members and students about the impact of mobile technology to support smart classes which are equipped with E-podium, a device that controls all the classrooms' components by an internal control unit along with special software [122]. The results indicate students are very much in favor of the use of E-podium and manifest expectations of regular up-to-date of the course materials to say the least.
- *Perceptions* are explored in Iranian universities, where users claim their perceptions about the mediating effect of usability towards use of m-learning, usefulness, subjective norm, image, innovativeness, individual mobility, absorptive capacity, and self-efficacy on user intention and satisfaction [123]. The study asserts those issues represent significant effects- that positively affect users' actual use.
- *Adequacy and perceptions* are measured from a sample of teachers and students to determine their attitude concerning m-learning use [124]. The conclusions reveal that both, teachers and students, want to use m-learning in education. Even though, their perceptions are positive, their m-learning adequacy levels are not sufficient.
- *Expectations and views* are measured as the learners' attitude towards the use of tablet PC [125]. So the "Computer attitude measure for young students" test [126] was applied to high school students and interviews were conducted with teachers. The findings indicate students have a positive attitude toward tablet PC.
- *Engagement* was observed in children interacting in small groups who use a story-making iPad app, called "Our Story", and other app software [127]. The evaluation was categorized by means of the Bangert-Drowns and Pyke's taxonomy [128]. The results unveil the quality of children's individual engagement was higher with the iPad app in contrast to their engagement with other app software.
- *Engagement* is observed in adults at lifelong learning settings, who in spite of being unfamiliar with m-learning technology, they took a literature course where users created routes of geolocation questions about a fiction book they were reading and answered them in real locations [129]. Outcomes show that adult's m-learning acceptance improved as their anxiety around use of technologies diminished.
- *Motivation* is investigated in [130] from the adults' learner perspective by means of the uses and gratification theory [131] to understand what motivates m-learning adoption. The findings suggest that adult learners' intention to use

m-learning is influenced by their cognitive, affective, and social needs through attitude.

- *Value* is analyzed in the light of: What do learners value as learning and how does this value relate to the use of mobile technology? [132]. A study, based on stimulated recall [133] and photo eliciting [134], involved adolescents who took photos of what they considered learning situations. The findings show that when the learners advice a special value for them, then they label the situation as learning.
- *Acceptance* is evaluated in addition to incidence and use of digital mobile devices among undergraduates [135]. Some findings correspond to diverse factors and variables that condition and favor the use of digital mobile devices in the university across three key areas: teaching model, area of study, and generic competencies.
- *Conceptions* about of context-aware u-learning are analyzed in [136] using a phenomenographic method [137]. The examination unveiled five categories of conceptions of u-learning, such as: the application of technology, the platform for attaining information, a timely guide, increase of knowledge, and active learning.
- *Reflection* is tackled in [138] by mobile notifications that foster reflective practice on meta-learning by two practices; (a) offering to adolescents a daily reflection and reporting exercise about their learning experience during the day; (b) inviting adults to read an eBook on energy-efficient driving for one hour. During the trial, subjects received mobile notifications inviting them to reflect in-action. The results from the first practice assert: students do not have a habit of seeing themselves as learners nor develop awareness of their activity at school. Whereas, the second explores the effects of diverse notification types on knowledge gain and motivation.

Education. In spite of MUP-Learning is concerned with education, a part of the research carried out focus on specific elements that facilitate teaching as well as learning as the following sample of works reveals in Table 4.

Diverse fields. Specific professional domains are also target of MUP-Learning applications, where specialized domain knowledge and skills support the development of a given competence as the following sample of works shows:

Messaging. A couple of works are oriented to examine the performance and influence of different messaging methods in m-learning sceneries. The first corresponds to [146], which examines the media richness of various message delivery methods in an m-learning environment based on media richness theory [147]. As a result, four factors (e.g., timeliness, richness, accuracy, and adaptability) are identified to evaluate a content in respect to the media richness among short message services, Email, and really simple syndication. One finding reveals that really simple syndication is more appropriate for supporting various front-end mobile devices to access and present the content in an m-learning environment.

Whereas the second work is published in [148] that analyzes how mobile instant messaging, personal computer-based instant messaging, and the bulletin board

Table 4 MUP-Learning sample oriented to educational purposes

Tool	Scope	Nature
[139]	Learning pills	Pills are simple exercises that summarize some of the key concepts explained in class in order to promote reflection and self-study. Authors assert that using mobile devices in class by means of contextualized learning pills improves learner class attendance, performance, and motivational patterns. Pills are automatically sent to students' mobile phone just after the related concept has been presented. It is concluded that students enjoy using mobile devices for learning, and they do that so even more at the end of the course than at the beginning
[140]	Integrating mobile phones into science teaching	The work shares an experience about a workshop series to train teachers on the use of m-phones in science teaching. The results show the educational potential of m-phones, in learning how to use them in teaching and learning, in changing their attitudes towards the use of m-phones, and in sharing skills and knowledge to m-phone applications
[141]	Brainwave activity	It is compared clicker technology against mobile polling and the just-in-time teaching strategy in how these methods affect students' anxiety, self-efficacy, engagement, attention, performance, and relaxation as indicated by brainwave activity. The results reveal that mobile polling along with the just-in-time teaching strategy and in-class polls reduce graduate students' anxiety, improve student results, and increase students' attention during polling
[142]	Instructional design processes	The study compares the difficulties that novice instructional designers experience during instructional design processes for mobile devices and desktop computers through the ADDIE (Analyze, Design, Develop, Implement, and Evaluate) model [143]. The results claim that the difficulties in internal design, production, and front-end analysis were different in terms of the Internet connection in personal mobile devices
[144]	Collaboration	The work examines how learners explore an environment as part of a simulated security guard training. Authors depict how social flow in a CSCL space might sketch out what triggers an optimal learning experience in collaboration, and what can be further achieved in a CSCL experience. They conclude that m-CSCL might prompt more knowledge generation and extra learning tasks by fostering greater motivation than other learning environments

(continued)

Table 4 (continued)

Tool	Scope	Nature
[145]	Couple games	Authors build an iPhone game for transmitting knowledge as part of multiculturalism, solidarity, and tolerance following learning theories, design principles, objectives and competences of the Spanish law. The game demands the collaboration of two children interacting at the same time in order each player proceeds to the next step. They also study whether the iPhone game has better learning outcomes than games based on consoles or computers

system exert on collaborative learning processes and outcomes. The work reveals that the mobile instant messaging exhibits better teamwork than the others groups; but, it demonstrates lower task work than both groups, bulletin board system and the personal computer-based instant messaging.

Arts and architecture. Both disciplines, arts and architecture, are the result of the inspiration for the fine arts devoted to manifest the most intrinsically feelings and thoughts of the human beings. In this section four works are summarized as a sample of the creativity supported by m-learning applications.

The first corresponds to [149], which explores AR in the visualization of three-dimension models and the presentation of architectural projects. The goal is to assess the feasibility of using AR on mobile devices in educational settings and study the relationship between usability of the tool, student participation, and improvement in academic performance after using AR. The evidences show that the use of mobile devices in the classroom is highly correlated with motivation, and there is a significant correlation with academic achievement.

As for the second work, [150] develops a phenomenographic examination on the use of iPads among art and design students. The study provided a student-centered perspective and the qualitative analysis revealed varied perceptions. The interpretation of the results uncover that there are a number of challenges and obstacles in embedding the use of iPads, as well as it cannot be claimed that students overall found that the iPad contributed in a significant manner towards their learning.

Another instance of arts work is described by Martin and Ertzberger [151], where the effects of here and now m-learning on student achievement and attitude are examined. In the trial, undergraduate students enrolled in preserve instructional design and instructional technology courses delivered through computers and iPad/iPod. The outcomes unveil that computer users achieved positive scores on the posttest, whilst the iPad/iPod learners had positive attitudes.

In the meantime, [152] analyzes the effects of sharing a mobile device within groups and their performance at indoor location-based learning activity. In order to explore such effects, the authors design a game, called "QuesTInSitu: The Game", to support a learning activity in a contemporary art museum. The results uncover that carrying the device does have a significant positive impact in their performance.

Workplace. Mobile technology is also used in workplaces to support diverse sorts of formal and informal learning as the following sample of approaches assert. The first is introduced by Gu et al. [153], who analyzed the impact of mobile Web 2.0 applications on informal learning in the workplace. Thus they examined employees' learning experiences with a mobile App to identify changes in participants' learning behaviors by analyzing each participant's use of an App integrating popular mobile Web 2.0 applications. They conclude that mobile devices with Web 2.0 applications were shown to be effective aids for informal learning in the workplace, as well as new learning habits can be formed and will benefit learners in long term.

By the side of Fuller and Joynes [154], they wonder: Should m-learning be compulsory for preparing students for learning in the workplace? So, they analyze how Medical School students retrieve mobile-based resources to support their learning and assessments as part of clinical activity in placement settings. Authors claim that educators should be focusing on developing mobile learning in curricula that is comprehensive, sustainable, meaningful, and compulsory, in order to prepare students for accessing and using such resources in their working lives.

Regarding to Diaconita et al. [155], they focus on how professionals keeping up with the rapid development in their fields, particularly when the new information is meant to be used for problem solving. Thus, they assert collaborative approaches (e.g., forums and question and answering) only partly mitigate this problem in cases where time constraints and user mobility are of no crucial concern. In consequence, authors present an enhancement to question and answering applications which relies on user activity detection and question forwarding towards the available users.

5 Discussion

Once a profile of the MUP-Learning works has been stated, in this section a quantitative analysis is provided as well as a perspective of the field is shaped. First, a sample of statistic and data mining outcomes is highlighted. Later on, several findings are unveiled to discover interesting correlations. Finally, the exposition of the SWOT is outlined to acknowledge a perspective of the MUP-Learning field.

5.1 *Quantitative Analysis*

With the purpose to tailor a quantitative perspective of the MUP-Learning arena, the database of the patterns that characterize the works, proposed in Sect. 3.5, is exploited to estimate frequencies and correlations through two statistical and data mining processes as follows.

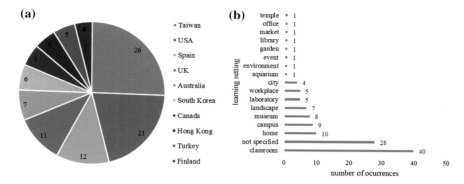

Fig. 2 Frequencies: **a** Top ten authors by country; **b** learning settings

Statistical Results. As a way to provide a partial image of the MUP-Learning arena, which could be useful to infer an overall idea of the whole field, a statistical analysis is presented in this subsection. Thus, several descriptive statistics and some correlations are estimated for the traits that compose the pattern to depict the sample of 105 collected works. As a sample of the achieved outcomes, the following results are described and illustrated as follows.

Concerning the most prolific countries in the development of MUP-Learning work, the Fig. 2a depicts the top ten frequencies of authors linked to a specific country. Whereas, another example appears in Fig. 2b where the topics related to learning settings and their number of occurrences are provided. Lastly, Fig. 3 shows a bar chart that represents the relation of frequencies derived from the subcategories pertaining to domain studies and the journal where those works were published.

As for the correlation between topics, an example appears in Fig. 4, where the journal $C\&E^1$ is related with the works published by the country *Taiwan*. The results illustrate a scatterplot estimated by an intercept of 0.23 and a slope of 0.04, as well as a correlation coefficient of 0.04 based on a confidence interval of 0.95, which means a low positive correlation. Even though the correlation is low, when a deeper analysis is made, the result is relevant due the sample size is 105 works!

Mining Outcomes. In order to find some correlations between the attributes that depict the MUP-Learning works, diverse data mining techniques are applied, such as clustering and association rules. The aim is to find out profiles that share similar characteristics that could be gathered into a specific aggrupation. Moreover, some criteria could be identified to constraint the relationship between diverse attributes.

As a result of the data mining processes applied to the database that contains the patterns that characterize the MUP-Learning works, several clusters and association rules were produced to unveil diverse patterns, as Figs. 5 and 6, respectively,

---

[1]The meaning of this journal as well as others that appears within figures and text in this section are identified in Appendix.

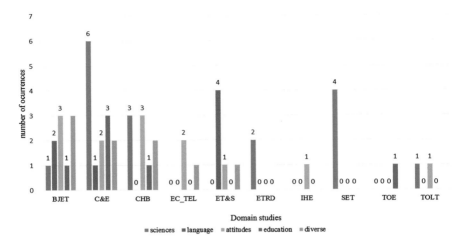

Fig. 3 Frequencies of subcategories of domain studies related to the journals

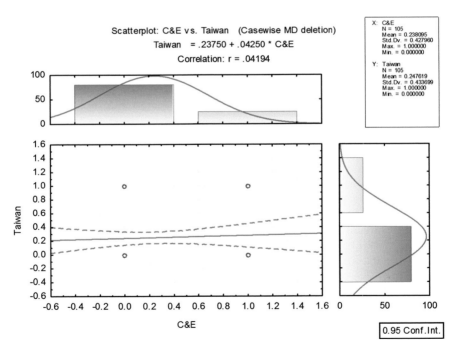

Fig. 4 Linear correlation between the journal C&E and the works coming from Taiwan

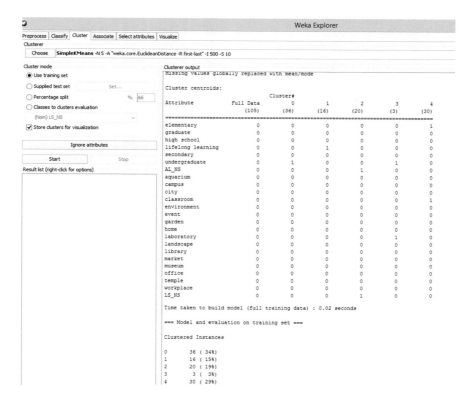

Fig. 5 Clusters that gathers instances of academic levels and learning settings

illustrate. For instance, Fig. 5 shows five clusters, where the one with the *id 3* gathers 22 profiles whose values for journal and country are *C&E* and *Taiwan,* respectively. Whereas, Fig. 6 uncovers four rules, where the number 3 asserts: from 23 patterns whose academic level is *elementary*, 15 of them hold the value *class-room* for the learning setting. More findings are stated in Sect. 5.2.

5.2 Findings

As part of the conceptual perspective of MUP-Learning arena, a sample of findings are uncovered in this subsection. Such findings are the result of the analysis made to the profile of related works introduced in Sect. 4, as well as the interpretation of the statistical and data mining outcomes already outlined.

- Regarding *domain studies* addressed by journals, Fig. 3 illustrates that C&E includes five subcategories identified in the *domain studies* category (i.e.,

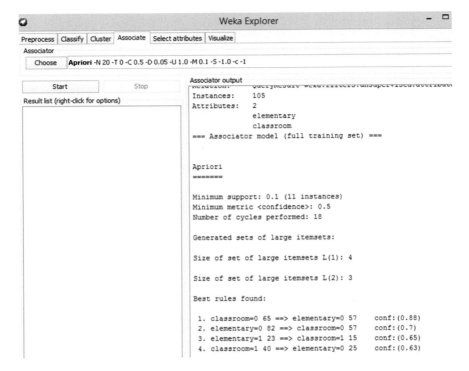

Fig. 6 Association rules that relate *elementary* academic level and *classroom* learning setting

sciences, language, attitudes, education and diverse). It represents the 27 % of the category.

- One cluster reveals the relation of the journal C&E with the subcategory of *sciences* pertaining to *domain studies* represents the 17 % of the sample.
- 9 % of the sample is joined in a cluster, which includes the journal C&E with works coming from Spain and UK.
- The journal BJET also embraces the five aforementioned subcategories, as Fig. 3 depicts it. This corresponds to the 19 % of the *domain studies* category.
- Moreover, BJET and the domain topic, *attitudes*, are included in a cluster corresponding to the 8 % of the sample. On the other hand, 6 % of the works are gathered in a cluster that contains the journal CHB and the domain topic *cognition*.
- In contrast, the journal ET&S gathers more works focused on *language learning* domain that reveal the 12 % of the category.
- Regarding works that introduce or use *frameworks*, these are gather in a cluster with the journal ET&S corresponding to a 12 % of the sample.
- On the subject of academic level, one cluster shows the relation between *elementary* school level and the country *Taiwan* which gathers 8 % of the whole sample. This corresponds to 31 % of the research production made in Taiwan.

- In addition, one cluster gathers works pertaining to *elementary* school level and *sciences* as domain of knowledge, corresponding to a 6 % of the sample.
- Concerning to *undergraduate* academic level one cluster reveals a relation with the instance *laboratory* of learning setting corresponding to 3 % of the sample. Furthermore, other mining case reflects one cluster where the domain topic *cognition* is addressed by *undergraduate* level in a 3 % of the sample.
- 10 % of the whole sample is gathered in a cluster containing *teaching* as domain topic and *education* as a subcategory of domain studies.
- According to the Fig. 2a, the country where there are the most occurrences of authors corresponds to *Taiwan* with 26 instances. It represents the 19 % of the total of authors' occurrences. Whilst, the second country heading the list is *USA* with 21 cases, which corresponds to the 15 % of the total.
- *Classroom* and *home* are the most frequent learning settings employed in MUP-Learning according to the surveyed works with a count of 40 and 10 instances respectively. As depicted in Fig. 2b, this represents the 40 % of the occurrences.
- In addition to learning settings, a mining case shows a cluster that links *classroom* and *home* with a 10 % of the whole sample.
- Other *indoor* learning settings (e.g., *museum, laboratory, workplace, aquarium,* etc) corresponds to the 18 % of the sample. Whilst, *outdoor* learning settings (e.g., *campus, landscape, city, garden...*) contain the 19 % of the surveyed works. It is worth mention that in a 23 % of the occurrences, the learning setting is not specified.

5.3 Analysis of Strengths, Weakness, Opportunities, and Threats

An essential part to tailor a viewpoint of the MUP-Learning field corresponds to the identification of its SWOT. Such aspects represent a reflection that highlights key issues that are worthy to be taken into account by developers and practitioners who are interested in carry out research, apply, and study MUP-Learning sceneries.

Strengths. Some aspects that reveal the fortress of the MUP-Learning, which should be considered to ground future development are presented next:

- A growing number of institutions and researchers are being incorporated and dedicated to build MUP-Learning systems.
- The studies on MUP-Learning suggest the students' willingness to accept and use such paradigm.
- Technological infrastructure and devices are cheap and affordable for all, people and academic institutions.
- The enhancements in technology and devices constantly provide better features that enhance the scope of the functionalities.

- The convenience and advantage of MUP-Learning platforms to access educational content anytime, anywhere, and anyway.
- The ease to perform observations and activities at indoor and outdoor locations as well as field trips tend to revolutionize the provision of educational services.

Weakness. As for the MUP-Learning issues that demand attention in order to overcome current limited value, the most significant are identified as follows:

- There is no standard definition for MUP-Learning, not to mention the MUP-Learning environment settings.
- MUP-Learning is partially applied to a subject or unit of a particular course.
- A few researches have been oriented to lifelong learning.
- Lack of well accepted software engineering models, frameworks, and platforms to easily develop MUP-Learning applications.
- Most of the work is concentrated in just one country, Taiwan.
- Research oriented to young children and seniors learners is scarce.

Opportunities. Like a source of motivation for researchers who are partisan or are considering being involved in MUP-Learning, several chances are introduced next:

- Definition of a robust baseline that encapsulates the theoretical foundations and best practices of the MUP-Learning arena.
- Promotion and development of personalized, adaptive, and reusable learning objects in MUP-Learning applications.
- Increment of MUP-Learning practices for situated, immersive, and meaningful courses and academic programs.
- Support research lines that aim the use of MUP-Learning in early childhood, seniors, and lifelong learning users.
- Development of tools oriented to aid learners with special educational needs.
- Spread of MUP-Learning sceneries in all academic levels.
- Encouragement of the collaboration between educational and computer science researchers, as well as experts of others disciplines (e.g., pedagogy, psychology, sociology, brain sciences, communication sciences, etc.) to build a holistic baseline.

Threats. Lastly, the challenges to be tackled by MUP-Learning developers interested in making an incursion correspond to:

- Keeping up the pace of the changing and emergent technologies.
- Government's unawareness about the benefits of implementing appropriate settings for MUP-Learning in the educational curricula.
- Insufficient budget to implement MUP-Learning settings in schools.
- Resistance to the change that usually functionaries, academics, and users of enhanced technology learning manifest.

- Diversity of learners' attitudes and willingness to deal with innovative MUP-Learning systems, where usually adults and seniors are less handy to deal with.
- Excessive cognitive load produced on the users' mind as a result of the diversity of stimuli that simultaneously affect the cognitive activity oriented to learn.

6   Conclusions

This chapter presents a perspective of the work accomplished in three learning research lines: mobile, ubiquitous, and pervasive learning, where a collection of 105 recent works has been classified according to the proposed taxonomy and pattern respectively. In this context, the statistical and data mining analysis reveal a vision of the MUP-Learning field. Thus, in order to conclude the chapter three topics are outlined. The first one highlights the future work, whereas the second identify diverse trends; whilst, the third provides the responses to the research questions.

Future Work. As part of the future work to be fulfilled, the design and promotion of MUP-Learning environments are strategic to intervene as a technological agent to build affordable teaching–learning environments with the purpose to improve education in all educational levels. As for the authors, we consider to develop a prototype to implement a cybernetic method to deal with cognitive load and enhance learning in MUP-Learning environments by means of SRL and metacognitive strategies.

Trends. Some of the most relevant trends are stated as follows: (1) In spite of the growth of works focused on undergraduate academic level, the elementary school remains as the most studied academic level; so, this means that new generations are being taught by means of MUP-Learning systems. (2) The incorporation of features and sensors within the devices and learning environments provides a valuable support for context-aware activities; thus, new educational environments claim for deploying this kind of settings. (3) The collaboration of experts from diverse disciplines should be considered in order to design a suitable and enriched learning experience to students; therefore, a multidisciplinary view is needed to develop holistic MUP-Learning approaches.

Response to Research Questions. As the final part of the work, a response is given for the research questions, stated in Sect. 3.1, that aimed the review. Thus, the answers are presented according to the order of the questions as follows:

1. The basics of the field are outlined in Sect. 2, where the background of educational systems is sketched, a series of definitions is pointed out, and a profile of prior surveys is drawn.
2. A taxonomy is proposed in Sect. 3.4 to classify the MUP-Learning works from two perspectives conceptual and empirical labor, and domain studies, which embrace specific subcategories to organize the sample of works.

3. A proposal for describing the MUP-Learning works is illustrated in Sect. 3.5 as a pattern that contains seven relevant traits, as well as its instantiation is illustrated in the Table A.1 of the Appendix.
4. A sample of the work currently performed is shown in Sect. 4, where a profile is written for each MUP-Learning work, and the sample of works is edited according to the sequence of categories and subcategories that compose the taxonomy. Also, the pattern that depicts such works is presented in Appendix.
5. A sample of the most frequent topics addressed in the MUP-Learning field is highlighted as a consequence of the statistical and data mining processes described in Sect. 5.1, where descriptive and correlational statistics, clusters, and association rules are unveiled and pictured through diverse figures.
6. Diverse relevant issues of the MUP-Learning field are claimed in the findings uncovered in Sect. 5.2, as well as the analysis of SWOT. Inclusive, the prior topics presented in this section, conclusions, add more highlights.

Acknowledgments The first author gives testimony of the strength given by his Father, Brother Jesus and Helper, as part of the research projects of World Outreach Light to the Nations Ministries (WOLNM). Moreover, this work holds a partial support from grants: CONACYT-SNI-36453, IPN-COFAA-SIBE-ID: 9020/2015-2016, IPN-SIP-EDI-848-14, IPN-SIP-20150910, CONACYT 289763, and IPN-SIP-BEIFI-20150910. IPN-SIP-20150910, CONACYT 264215, CONACYT 289763, and IPN-SIP-BEIFI-597.

Appendix

The pattern that characterizes the main attributes of the sample of MUP-Learning works, described in Sect. 3.5, is edited in Table A.1, where seven key traits are shown. The series of patterns is presented according to the hierarchy sketched in the Taxonomy to classify MUP-Learning works, illustrated in Table 1, as well as the profile stated for each related work along Sect. 4.

Inclusive, the name of each category, subcategory, and topic as well as their respective count of related works is presented as a header inside of the table body. However, due the lack of space, diverse acronyms are used to facilitate the edition of several common values, according to next definitions:

- Journal:

  - BJET: British Journal of Educational Technology.
  - C&E: Computers & Education.
  - CHB: Computers in Human Behavior.
  - EC_TEL: European Conference on Technology Enhanced Learning.
  - ESWA: Expert Systems with Applications.
  - ET&S: Educational Technology & Society.
  - ETRD: Education Tech Research Development.
  - ICALT: International Conference on Advanced Learning Technologies.

Table A.1 Description of the sample works according to the pattern to characterize MUP-Learning works

Citation	Year	Source	Country	Academic level	Learning setting	Domain
Sect. 4.1 Conceptual and empirical: 53 works						
Models: 13 works						
[29]	2014	BJET	Taiwan	Elementary	Campus	Natural science
[30]	2013	C&E	Australia	Undergraduate	N/S	Cognition
[31]	2012	CHB	USA, Singapore	Elementary	Classroom	Cognition
[32]	2014	CHB	USA, Australia	Undergraduate	N/S	ICT
[34]	2015	CHB	Taiwan	High school, undergraduate	Museum, aquarium	Environment
[36]	2014	ESWA	Brazil	Undergraduate	Campus	Adaptivity
[37]	2014	ET&S	Taiwan	Elementary	Temple	Culture
[38]	2011	ETRD	USA	N/S	Classroom, museum, library	ICT
[39]	2014	SET	Turkey, USA	Undergraduate	N/S	Environment
[41]	2013	C&E	Israel, USA	Lifelong learning	Office	Road safety
[44]	2014	SET	USA	Elementary	Landscape	Environment
[46]	2012	TOLT	Australia, Saudi Arabia	N/S	N/S	Adaptivity
[48]	2015	TOLT	Israel	Undergraduate	Classroom, museum, home	Arts
Frameworks: 17 works						
[49]	2012	BJET	USA	N/S	N/S	Teaching
[50]	2013	BJET	Australia	Secondary	Classroom, home	Exact science, natural science
[52]	2013	BJET	Hong Kong	N/S	N/S	Teaching
[53]	2010	C&E	USA, Chile	Elementary	Classroom	Exact science

(continued)

Table A.1 (continued)

Citation	Year	Source	Country	Academic level	Learning setting	Domain
[55]	2014	C&E	USA, Canada	Secondary	Classroom	Teaching
[57]	2015	CHB	Italy	High school	City	Arts
[58]	2010	ET&S	UK, Germany	Undergraduate	N/S	Support
[59]	2013	ET&S	Japan, Hong Kong, Taiwan	Graduate	Classroom	ICT
[61]	2013	ET&S	Singapore	Elementary	Classroom, campus	Language
[62]	2014	ET&S	Singapore	Elementary	Classroom	Environment
[63]	2014	ET&S	Hong Kong	Elementary	Classroom	Natural science
[64]	2014	ICALT	Germany	N/S	N/S	Adaptivity
[65]	2015	TOLT	USA	Lifelong learning	N/S	ICT
[66]	2014	SLE	Taiwan	N/S	N/S	Learning
[67]	2013	ULET	China	N/S	Classroom	Classroom
[68]	2011	C&E	Taiwan	Elementary	Campus	Language
[69]	2014	UMUAI	Taiwan	N/S	Campus	Natural science
Tools: 6 works						
[70]	2010	C&E	Spain	High school, undergraduate	Classroom	Learning
[71]	2014	C&E	Taiwan	Elementary	Classroom, landscape	Physical science
[72]	2014	EC_TEL	Netherlands	Lifelong learning	Environment	Cognition
[73]	2014	EC_TEL	Spain	N/S	Classroom	Special education
[75]	2015	ULET	Canada, Taiwan	N/S	N/S	Decision-making
[76]	2015	ETRD	Malaysia	Secondary	Classroom, home	Teaching

(continued)

Table A.1 (continued)

Citation	Year	Source	Country	Academic level	Learning setting	Domain
Approaches: 17 works						
[77]	2013	C&E	Spain	Elementary	Classroom	Special education
[78]	2013	C&E	UK	N/S	Classroom, home	Special education
[80]	2012	C&E	UK	Secondary	Classroom, home	Special education
[81]	2013	C&E	Finland, Taiwan, Canada	Lifelong learning	Workplace	Support
[82]	2014	CHB	Taiwan	Undergraduate	Classroom	Classroom
[83]	2014	EC_TEL	Italy	Graduate	N/S	Adaptivity
[84]	2011	ESWA	Taiwan	Lifelong learning	N/S	Adaptivity
[85]	2012	ESWA	South Korea	Undergraduate	Campus	Decision-making
[86]	2014	ESWA	Brazil	Undergraduate	Campus	Recommendation
[87]	2015	ET&S	South Korea, Finland	Undergraduate, elementary, lifelong learning	Event	Health
[88]	2011	KMEL	Canada, Taiwan	Graduate	Workplace	Training
[89]	2013	RPTEL	Japan	Undergraduate	City	Support
[90]	2015	ULET	Canada, China, Finland	N/S	N/S	Language
[91]	2012	ET&S	Taiwan	Lifelong learning	City	Road safety
[92]	2012	ET&S	Taiwan	Lifelong learning	Museum	Recommendation
[93]	2014	ET&S	UK	Lifelong learning	Landscape	Natural science
[94]	2014	ET&S	Taiwan	Undergraduate	Museum	Arts

(continued)

Table A.1 (continued)

Citation	Year	Source	Country	Academic level	Learning setting	Domain
4.2 Domain Studies: 52 works						
Sciences: 17 works						
[95]	2012	BJET	UK	Secondary	Museum	Attitudes
[96]	2010	C&E	Germany	Lifelong learning	Landscape	Environment
[97]	2013	C&E	UK	Lifelong learning	Classroom, landscape	Physical science natural science
[98]	2014	C&E	Hong Kong	Elementary	Classroom, home, market	Natural science
[99]	2014	SET	Taiwan, Australia, China	Undergraduate	N/S	Physical science
[100]	2015	TOLT	Sweden	Elementary	Landscape	Natural science
[102]	2014	C&E	Taiwan	Elementary	Laboratory	Physical science natural science
[103]	2013	CHB	Taiwan	Undergraduate	Laboratory	Cognition
[104]	2013	C&E	Spain	Elementary	Classroom	Natural science
[105]	2015	ETRD	Zambia, Finland	Elementary	Classroom	Cognition
[106]	2015	ETRD	USA	Lifelong learning	Landscape	Natural science
[107]	2014	SET	Australia	High school	Classroom	Natural science
[108]	2014	C&E	Taiwan, Netherlands, Australia	Elementary	Garden	Natural science
[109]	2015	SET	USA	Secondary, high school	Classroom	Teaching
[110]	2013	SET	USA	Undergraduate	Laboratory	Natural science
[111]	2014	CHB	USA	High school	Classroom	Physical science
[112]	2013	CHB	South Korea, USA	Undergraduate	N/S	Physical science

(continued)

Table A.1 (continued)

Citation	Year	Source	Country	Academic level	Learning setting	Domain
Sect. 4.2 Domain Studies: 52 works						
Language learning: 7 works						
[113]	2013	ET&S	Taiwan	Elementary	Classroom	Language
[114]	2013	ET&S	Singapore, Taiwan, Croatia	Elementary	Classroom	Language
[115]	2014	ET&S	USA	Elementary, secondary	Classroom, home	Language
[116]	2012	BJET	USA, South Korea	Secondary	N/S	Language
[117]	2012	BJET	Taiwan, USA	Undergraduate	N/S	Language
[118]	2013	C&E	Sweden	High school	N/S	Language
[119]	2010	ET&S	Netherlands	N/S	Laboratory	Language
Attitudes: 13 works						
[120]	2014	BJET	Turkey	Undergraduate	N/S	Attitudes
[121]	2014	C&E	USA	High school, secondary, elementary	N/S	Attitudes
[122]	2014	CHB	Saudi Arabia	Graduate, undergraduate	N/S	Attitudes
[123]	2015	CHB	Iran	Undergraduate	N/S	Attitudes
[124]	2015	BJET	Turkey	Secondary	N/S	Attitudes
[125]	2014	CHB	Turkey	High school	Classroom, home	Attitudes
[127]	2013	C&E	UK, Spain	N/S	Classroom	Attitudes
[129]	2013	EC_TEL	UK, Spain	Lifelong learning	City	Arts
[130]	2013	BJET	Malaysia	Graduate	N/S	Attitudes
[132]	2014	EC_TEL	Sweden	Elementary	Home, classroom	Learning
[135]	2015	ET&S	Spain	Undergraduate, graduate	N/S	Attitudes
[136]	2011	IHE	Taiwan	Undergraduate	Museum	Cognition
[138]	2015	TOLT	Spain, Netherlands	Secondary, lifelong learning	Laboratory	Cognition

(continued)

Table A.1 (continued)

Citation	Year	Source	Country	Academic level	Learning setting	Domain
Education: 6 works						
[139]	2011	TOE	Spain	Undergraduate	Classroom	ICT
[140]	2015	BJET	Sri Lanka, UK	Lifelong learning	Classroom	Teaching
[141]	2014	C&E	Taiwan	Graduate, undergraduate	Classroom	ICT
[142]	2014	CHB	Turkey	Undergraduate	Classroom	Teaching
[144]	2014	C&E	South Korea	Graduate	Campus	Training
[145]	2013	C&E	Spain	Elementary	Classroom	Culture
Diverse fields: 9 works						
Messaging: 2 works						
[146]	2010	C&E	Taiwan	Undergraduate	N/S	ICT
[148]	2014	ET&S	South Korea, USA	Undergraduate	Classroom, home	ICT
Arts and architecture: 4 works						
[149]	2014	CHB	Spain	Undergraduate	Classroom	Architecture
[150]	2015	BJET	UK, Cyprus	Undergraduate	N/S	Arts
[151]	2013	C&E	USA	Undergraduate	Classroom, campus	Arts
[152]	2015	CHB	Spain	Secondary	Museum	Arts
Workplace: 3 works						
[153]	2014	BJET	Hong Kong	Lifelong learning	Workplace	Support
[154]	2015	BJET	UK	Undergraduate	Workplace	Health
[155]	2014	EC_TEL	Germany	Lifelong learning	Workplace	Support

- IHE: Internet and Higher Education.
- KMEL: Knowledge Management & E-Learning: An International Journal.
- RPTEL: Research and Practice in Technology Enhanced Learning.
- SET: Journal of Science and Technology.
- SLE: Smart Learning Environments.
- TOE: IEEE Transactions on Education.
- TOLT: IEEE Transactions on Learning Technologies.
- ULET: Ubiquitous Learning Environments and Technologies.
- UMUAI: User Modeling and User-Adapted Interaction.

- Country: UK: United Kingdom; USA: United States of America.
- Academic level: N/S: not specified.
- Learning setting: N/S: not specified.
- Domain: ICT: information–communication technologies.

References

1. McDonald, J.K., Yanchar, S.C., Osguthorpe, R.T.: Learning from programmed instruction: examining implications for modern instructional technology. Educ. Tech. Res. Dev. **53**(2), 84–98 (2005)
2. Bratthall, D.: Programmed self-instruction in oral hygiene. J. Periodontal Res. **2**(3), 207–214 (2006)
3. S.L.A: Simple apparatus which gives tests and scores and teaches. J. Sch. Soc. **23**(586), 373–376 (1926)
4. Petrina, S.: Sidney Pressey and the automation of education, 1924–1934. J. Technol. Cult. **45** (2), 305–330 (2004)
5. Hunter B.: Learning, teaching, and building knowledge: a forty-year quest for online learning communities. In: Online Learning: Personal Reflections on the Transformation of Education, pp. 163–193. Educational Technology Publications, New Jersey (2005)
6. Sleeman, D., Brown, J. (eds.): Intelligent Tutoring System. Academic Press, London (1982)
7. Psotka, J., Massey, L.D., Mutter, S.A. (eds.): Intelligent Tutoring Systems: Lessons Learned. Lawrence Erlbaum Associates Inc., New Jersey (1989)
8. Mulwa, C., Lawless, S., Sharp, M., Arnedillo-Sanchez, I., Wade, V.: Adaptive educational hypermedia systems in technology enhanced learning: a literature review. In: Proceedings of the Conference on Information Technology Education, pp. 73–84 (2010)
9. Zurita, G., Nussbaum, M.: Computer supported collaborative learning using wirelessly interconnected handheld computers. Comput. Educ. **42**(3), 289–314 (2004)
10. Sclater, N.: Web 2.0, personal learning environments, and the future of learning management systems. Res. Bull. **13**(13), 1–13 (2008)
11. Graf, S., Liu, T.C., Kinshuk Chen, N.S., Yang, S.J.H.: Learning styles and cognitive traits— their relationship and its benefits in web-based educational systems. Comput. Hum. Behav. **25**(6), 1280–1289 (2009)
12. Desmarais, M.C., Baker, R.S.: A review of recent advances in learner and skill modeling in intelligent learning environments. User Model. User-Adap. Inter. **22**(1–2), 9–38 (2012)
13. Kukulska-Hulme, A., Traxler, J. (eds.): Mobile Learning: A Handbook for Educators and Trainers. Psychology Press, East Sussex (2005)
14. Hung, H.T., Yuen, S.C.Y.: Educational use of social networking technology in higher education. Teach. High. Educ. **15**(6), 703–714 (2010)

15. Laine, T., Joy, M.: Survey on context-aware pervasive learning environments. Int. J. Interact. Mobile Technol. **3**(1), 70–76 (2009)
16. Peña-Ayala, A. (ed.): Educational Data Mining: Applications and Trends. Springer, Heidelberg (2014)
17. Chen, G.D., Chang, C.K., Wang, C.Y.: Ubiquitous learning website: Scaffold learners by mobile devices with information-aware techniques. Comput. Educ. **50**(1), 77–90 (2008)
18. Taniar, D. (ed.): Mobile Computing: Concepts, Methodologies, Tools, and Applications, vol. 1, pp. xiii. IGI Global, Hershey (2008)
19. UNESCO. http://www.unesco.org/new/en/unesco/themes/icts/m4ed/
20. Lucke, U., Rensing, C.: A survey on pervasive education. Pervasive Mobile Comput. **14**, 13–16 (2014)
21. Weiser, M.: The computer for the twenty-first century. Sci. Am. IEEE Xplore **265**(3), 94–110 (1991)
22. Peng, H., Su, Y., Chou, C., Tsai, C.: Ubiquitous knowledge construction: mobile learning redefined and a conceptual framework. Innov. Educ. Teach. Int. **46**(2), 171–183 (2009)
23. Hwang, G.J., Tsai, C.C., Yang, S.J.H.: Criteria, strategies and research Issues of context aware ubiquitous learning. Educ. Technol. Soc. **11**(2), 81–91 (2008)
24. Alsiyami, A.: A policy language definition for provenance in pervasive computing. Ph.D Dissertation. University of Sussex, Brighton (2012)
25. Sherimon, P.C., Reshmy, K.: Towards pervasive mobile learning—the vision of 21st century. Procedia Soc. Behav. Sci. **15**, 3067–3073 (2011)
26. Song, Y.: Methodological issues in mobile computer-supported collaborative learning (mCSCL): what methods, what to measure and when to measure? Educ. Technol. Soc. **17**(4), 33–48 (2014)
27. Gilman, E., Sanchez, I., Cortes, M., Riekki, J.: Towards user support in ubiquitous learning system. IEEE Trans. Learn. Technol. **8**(1), 55–68 (2015)
28. Pachler, N., Bachmair, B., Cook, J.: Mobile Learning: Structures, Agency, Practices. pp. 155–171 Springer, New York (2010)
29. Huang, Y.M., Chiu, P.-S.: The effectiveness of a meaningful learning-based evaluation model for context-aware mobile learning. Br. J. Educ. Technol. **46**(2), 437–447 (2015)
30. Lee, C.B.: Exploring the relationship between intention to use mobile phone as a visualization tool and regulation of cognition. Comput. Educ. **60**(1), 138–147 (2013)
31. Sha, L., Loo, C.K., Chen, W., Seow, P., Wong, L.H.: Recognizing and measuring self-regulated learning in a mobile learning environment. Comput. Hum. Behav. **28**(2), 718–728 (2012)
32. Mills, L.A., Knezek, G., Khaddage, F.: Information seeking, information sharing, and going mobile: three bridges to informal learning. Comput. Hum. Behav. **32**, 324–334 (2014)
33. Maniotes, L.K.: The transformative power of literary third space. PhD. Dissertation. University of Colorado, Colorado (2005)
34. Liao, Y.W., Huang, Y.M., Chen, H.C., Huang, S.H.: Exploring the antecedents of collaborative learning performance over social networking sites in a ubiquitous learning context. Comput. Hum. Behav. **43**, 313–323 (2015)
35. Davis, F.D., Bagozzi, R.P., Warshaw, P.R.: User acceptance of computer technology: a comparison of two theoretical models. Manage. Sci. **35**(8), 982–1003 (1989)
36. Wagner, A., Barbosa, J.L.V., Barbosa, D.N.F.: A model for profile management app to ubiquitous learning environments. Expert Syst. Appl. **41**(4), 2023–2034 (2014)
37. Chen, C.P., Shih, J.L., Ma, Y.C.: Using instructional pervasive game for school children's cultural learning. Educ. Technol. Soc. **17**(2), 169–182 (2014)
38. Arnone, M.P., Small, R.V., Chauncey, S.A., McKenna, P.: Curiosity, interest and engagement in technology-pervasive learning environments: a new research agenda. Educ. Technol. Res. Dev. **59**(2), 181–198 (2011)
39. Çalik, M., Özsevgec, T., Ebenezer, J., Artun, H., Küçük, Z.: Effects of 'environmental chemistry' elective course via technology-embedded scientific inquiry model on some variables. J. Sci. Educ. Technol. **23**, 412–430 (2014)

40. Ebenezer, J., Kaya, O.N., Ebenezer, D.L.: Engaging students in environmental research projects: perceptions of fluency with innovative technologies and levels of scientific inquiry abilities. J. Res. Sci. Teach. **48**(1), 94–116 (2011)
41. Reychav, I., Wu, D.: Exploring mobile tablet training for road safety: a uses and gratifications perspective. Comput. Educ. **71**, 43–55 (2014)
42. Palmgreen, P.: Uses and gratifications: a theoretical perspective. In: Bostrom, R.N. (ed.) Communication Yearbook, vol. 8, pp. 20–55. Sage, Beverly Hills (1984)
43. Fornell, C., Larcker, D.F.: Evaluating structural equation models with unobservable variables and measurement error. J. Mark. Res. **18**, 39–50 (1981)
44. Boyce, C.J., Mishra, C., Halverson, K.L., Thomas, A.K.: Getting students outside: using technology as a way to stimulate engagement. J. Sci. Educ. Technol. **23**, 815–826 (2014)
45. Roschelle, J.J.: Unlocking the learning value of wireless mobile devices. J. Comput. Assist. Learn. **19**, 260–272 (2003)
46. Al-Hmouz, A., Shen, J., Al-Hmouz, R., Yan, J.: Modeling and simulation of an adaptive neuro-fuzzy inference system (ANFIS) for mobile learning. IEEE Trans. Learn. Technol. **5**(3), 226–237 (2012)
47. Al-Hmouz, A., Shen, J., Yan, J., Al-Hmouz, R.: Enhanced learner model for adaptive mobile learning. In: Kotsis, G., Taniar, D., Pardede, E., Saleh, I., Khalil, I. (eds.) Proceeding 12th International Conference of Information Integration and Web-Based Applications and Services (IIWAS), pp. 781–784 (2010)
48. Kali, Y., Sagy, O., Kuflik, Mogilevsky, O., Maayan-Fanar, E.: Harnessing technology for promoting undergraduate art education: a novel model that streamlines learning between classroom, museum, and home. IEEE Trans. Learn. Technol. **8**(1), 72–84 (2015)
49. Wang, M., Shen, R.: Message design for mobile learning: learning theories, human cognition and design principles. Br. J. Educ. Technol. **43**(4), 561–575 (2012)
50. Ng, W., Nicholas, H.: A framework for sustainable mobile learning in schools. Br. J. Educ. Technol. **44**(5), 695–715 (2013)
51. Cisler, S.: Planning for sustainability: how to keep your ICT project running (schools online). http://geoinfo.uneca.org/sdiafrica/Reference/Ref6/Sustainabilit-booklet.doc
52. Kong, S.C., Song, Y.: A principle-based pedagogical design framework for developing constructivist learning in a seamless learning environment: a teacher development model for learning and teaching in digital classrooms. Br. J. Educ. Technol. **44**(6), E209–E212 (2013)
53. Roschelle, J., Rafanan, K., Estrella, G., Nussbaum, M., Claro, S.: From handheld collaborative tool to effective classroom module: embedding CSCL in a broader design framework. Comput. Educ. **55**(3), 1018–1026 (2010)
54. Bannan-Ritland, B.: The role of design in research: the integrative learning design framework. Educ. Res. **32**(1), 21–24 (2003)
55. Green, L.S., Hechter, R.P., Tysinger, P.D., Chassereau, K.D.: Mobile app selection for 5th through 12th grade science: the development of the MASS rubric. Comput. Educ. **75**, 65–71 (2014)
56. Kearney, M., Schuck, S., Burden, K., Aubusson, P.: Viewing mobile learning from a pedagogical perspective. Res. Learning Technol. **20**, 1–17 (2012)
57. Fulantelli, G., Taibi, D., Arrigo, M.: A framework to support educational decision making in mobile learning. Comput. Hum. Behav. **47**, 50–59 (2015)
58. Yau, J.Y.K., Joy, M., Dickert, S.: A mobile context-aware framework for managing learning schedules—data analysis from a diary study. Educ. Technol. Soc. **13**(3), 22–32 (2010)
59. Yin, C., Song, Y., Tabata, Y., Ogata, H., Hwang, G.J.: Developing and implementing a framework of participatory simulation for mobile learning using scaffolding. Educ. Technol. Soc. **16**(3), 137–150 (2013)
60. Kolb, D.A.: Experiential Learning: Experience as the Source of Learning and Development. Prentice Hall, Englewood Cliffs (1984)
61. Wong, L.H.: Analysis of students' after-school mobile-assisted artifact creation processes in a seamless language learning environment. Educ. Technol. Soc. **16**(2), 198–211 (2013)

62. Looi, C.K., Wong, L.H.: Implementing mobile learning curricula in schools: a programme of research from innovation to scaling. Educ. Technol. Soc. **17**(2), 72–84 (2014)
63. Kong, S.C., Song, Y.: The impact of a principle-based pedagogical design on inquiry-based learning in a seamless learning environment in Hong Kong. Educ. Technol. Soc. **17**(2), 127–141 (2014)
64. Moebert, T., Jank, H., Zender, R., Lucke, U.: A generalized approach for context-aware adaption in mobile e-learning settings. In: IEEE 14th International Conference on Advanced Learning Technologies, pp. 143–145 (2014)
65. Brinton, C., Rill, R., Ha, S., Chiang, M., Fellow, Smith, R., Ju, W.: Individualization for education at scale: MIIC design and preliminary evaluation. IEEE Trans. Learn. Technol. **8** (1), 72–84 (2015)
66. Hwang, G.J.: Definition, framework and research issues of smart learning environments-a context-aware ubiquitous learning perspective. Smart Learning Environ **1**(4), 1–14 (2014)
67. Huang, R., Hu, Y., S., Yang, J.: Improving learner experience in the technology rich classroom. In: Huang, R. (ed.) Lecture Notes in Educational Technology, Ubiquitous Learning Environments and Technologies, pp. 243–257. Springer, Berlin (2015)
68. Roschelle, J., Rafanan, K., Estrella, G., Nussbaum, M., Claro, S.: From handheld collaborative tool to effective classroom module: embedding CSCL in a broader design framework. Comput. Educ. **55**(3), 1018–1026 (2010)
69. Su, J.M., Tseng, S.S., Lin, H.Y., Chen, C.H.: A personalized learning content adaptation mechanism to meet diverse user needs in mobile learning environments. User Model User-Adapt. Interact. **21**(1–2), 5–49 (2011)
70. de-Marcos, L., Hilera, J.R., Barchino, R., Jiménez, L., Martínez, J.J., Gutiérrez, J.A., Gutiérrez, J.M., Otón, S.: An experiment for improving students performance in secondary and tertiary education by means of m-learning auto-assessment. Comput. Educ. **55**(3), 1069–1079 (2010)
71. Zhang, J., Sung, Y.T., Hou, H.T., Chang, K.E.: The development and evaluation of an augmented reality based armillary sphere for astronomical observation instruction. Comput. Educ. **73**, 178–188 (2014)
72. Tabuenca, B., Kalz, M., Specht, M.: Lifelong learning hub: a seamless tracking tool for mobile learning. In: Rensing, C., de Freitas, S., Muñoz-Merino, P.J. (eds.) EC-TEL 2014. LNCS 8719, pp. 534–537. Springer, Switzerland (2014)
73. Cabielles-Hernández, D., Pérez-Pérez, J.R., Paule-Ruiz, M.P., Álvarez-García, V.M., Fernández-Fernández, S.: dmTEA: mobile learning to aid in the diagnosis of autism spectrum disorders. In: Rensing, C., de Freitas, S., Muñoz-Merino, P.J. (eds.) EC-TEL 2014. LNCS 8719, pp. 29–41. Springer, Switzerland (2014)
74. Rivière, Á., Martos, J.: Tratamiento del autismo. Nuevas perspectivas. Instituto de Migraciones y Servicios Sociales, Madrid (1998)
75. Mottus, A., Graf, S., Chen, N.S.: Use of dashboards and visualization techniques to support teacher decision making. In: Huang, R. (ed.) Lecture Notes in Educational Technology, Ubiquitous Learning Environments and Technologies, pp. 181–199. Springer, Berlin (2015)
76. DeWitt, D., Alias, N., Siraj, S.: The design and development of a collaborative mLearning prototype for Malaysian secondary school science. Educ. Technol. Res. Dev. **62**, 461–480 (2014)
77. Fernández-López, A., Rodríguez-Fórtiz, M.J., Rodríguez-Almendros, M.L., Martínez-Segura, M.J.: Mobile learning technology based on iOS devices to support students with special education needs. Comput. Educ. **61**, 77–90 (2013)
78. Mintz, J.: Additional key factors mediating the use of a mobile technology tool designed to develop social and life skills in children with autism spectrum disorders: evaluation of the 2nd HANDS prototype. Comput. Educ. **63**, 17–27 (2013)
79. Fogg, B.J.: Persuasive Technology. Using Computers to Change What We Think and Do. Morgan Kaufman Publishers, San Francisco (2003)

80. Mintz, J., Branch, C., March, C., Lerman, S.: Key factors mediating the use of a mobile technology tool designed to develop social and life skills in children with autistic spectrum disorders. Comput. Educ. **58**(1), 53–62 (2012)
81. Yang, G., Chen, N.S., Kinshuk, Sutinen, E., Anderson, T., Wen, D.: The effectiveness of automatic text summarization in mobile learning contexts. Comput. Educ. **68**, 233–243 (2013)
82. Shen, C.W., Wu, Y.C., Lee, T.C.: Developing a NFC-equipped smart classroom: effects on attitudes toward computer science. Comput. Hum. Behav. **30**, 731–738 (2014)
83. Mercurio, M., Torre, I., Torsani, S.: Responsive web and adaptive web for open and ubiquitous learning. In: Rensing, C., de Freitas, S., Muñoz-Merino, P.J. (eds.) EC-TEL 2014. LNCS 8719, pp. 452–457. Springer, Switzerland (2014)
84. Wang, S.L., Wu, C.Y.: Application of context-aware and personalized recommendation to implement an adaptive ubiquitous learning system. Expert Syst. Appl. **38**(9), 10831–10838 (2011)
85. Lee, K.C., Cho, H.: Integration of general Bayesian network and ubiquitous decision support to provide context prediction capability. Expert Syst. Appl. **39**(5), 6116–6121 (2012)
86. de Neves, A.R., Carvalho, A.M.G., Ralha, C.G.: Agent-based architecture for context-aware and personalized event recommendation. Expert Syst. Appl. **41**(2), 563–573 (2014)
87. Laine, T.H., Islas Sedano, C.: Distributed pervasive worlds: the case of exergames. Educ. Technol. Soc. **18**(1), 50–66 (2015)
88. Lu, C., Chang, M., Kinshuk, Huang, E., Chen, C.W.: Usability of context-aware mobile educational game. Knowledge management and e-learning. Int. J. **3**(3), 448–475 (2011)
89. Mouri, K., Ogata, H., Li, M., Hou, B., Noriko, U., Liu, S.: Learning log navigator: supporting task-based learning using ubiquitous learning logs. Res. Pract. Technol. Enhanced Learning **8**(1), 117–128 (2013)
90. Wen, D., Gao, Y., Yang, G.: Semantic analysis-enhanced natural language interaction in ubiquitous learning. In: R. Huang (ed.) Ubiquitous Learning Environments and Technologies, Lecture Notes in Educational Technology, Springer, Heidelberg (2015)
91. Lan, Y.F., Huang, S.M.: Using mobile learning to improve the reflection: a case study of traffic violation. Educ. Technol. Soc. **15**(2), 179–193 (2012)
92. Huang, Y.M., Liu, C.H., Lee, C.Y., Huang, Y.M.: Designing a personalized guide recommendation system to mitigate information overload in museum learning. Educ. Technol. Soc. **15**(4), 150–166 (2012)
93. Scanlon, E., Woods, W., Clow, D.: Informal participation in science in the UK: identification, location and mobility with iSpot. Educ. Technol. Soc. **17**(2), 58–71 (2014)
94. Hou, H.T., Wu, S.Y., Lin, P.C., Sung, Y.T., Lin, J.W., Chang, K.E.: A blended mobile learning environment for museum learning. Educ. Technol. Soc. **17**(2), 207–218 (2014)
95. Charitonos, K., Blake, C., Scanlon, E., Jones, A.: Museum learning via social and mobile technologies: (How) can online interactions enhance the visitor experience? Br. J. Educ. Technol. **43**(5), 802–819 (2012)
96. Ruchter, M., Klar, B., Geiger, W.: Comparing the effects of mobile computers and traditional approaches in environmental education. Comput. Educ. **54**(4), 1054–1067 (2010)
97. Jones, A.C., Scalon, E., Clough, G.: Mobile learning: two case studies of supporting inquiry learning in informal and semiformal settings. Comput. Educ. **61**, 21–32 (2013)
98. Song, Y.: "Bring Your Own Device (BYOD)" for seamless science inquiry in a primary school. Comput. Educ. **74**, 50–60 (2014)
99. Wang, H.Y., Duh, H.B.L., Li, N., Lin, T.J., Tsai, C.C.: An investigation of university students' collaborative inquiry learning behaviors in an augmented reality simulation and a traditional simulation. J. Sci. Educ. Technol. **23**, 682–691 (2014)
100. Nouri, J., Cerratto-Pargman, T.: Characterizing learning mediated by mobile technologies: a cultural-historical activity theoretical analysis. IEEE Trans. Learn. Technol. **99**, 1–11 (2015)
101. Zurita, G., Nussbaum, M.: A conceptual framework based on Activity Theory for mobile CSCL. Br. J. Educ. Technol. **38**(2), 211–235 (2007)

102. Hong, J.H., Hwang, M.Y., Liu, M.C., Ho, H.Y., Chen, Y.L.: Using a "prediction–observation–explanation" inquiry model to enhance student interest and intention to continue science learning predicted by their Internet cognitive failure. Comput. Educ. **72**, 110–120 (2014)
103. Jou, M., Wang, J.: Ubiquitous tutoring in laboratories based on wireless sensor networks. Comput. Hum. Behav. **29**(2), 439–444 (2013)
104. Furió, D., González-Gancedo, S., Juan, M.C., Seguí, I., Costa, M.: The effects of the size and weight of a mobile device on an educational game. Comput. Educ. **64**, 24–41 (2013)
105. Jere-Folotiya, J., Chansa-Kabali, T., Munachaka, J.C., Sampa, F., Yalukanda, C., Westerholm, J., Richardson, U., Serpell, R., Lyytinen, H.: The effect of using a mobile literacy game to improve literacy levels of grade one students in Zambian schools. Educ. Tech. Res. Dev. **62**, 417–436 (2014)
106. Land, S.M., Zimmerman, H.T.: Socio-technical dimensions of an outdoor mobile learning environment: a three-phase design-based research investigation. Educ. Tech. Res. Dev. **63**, 229–255 (2015)
107. Incantalupo, L., Treagust, D.F., Koul, R.: Measuring student attitude and knowledge in technology-rich biology classrooms. J. Sci. Educ. Technol. **23**, 98–107 (2014)
108. Liu, T.C., Lin, Y.C., Paas, F.: Effects of prior knowledge on learning from different compositions of representations in a mobile learning environment. Comput. Educ. **72**, 328–338 (2014)
109. Shirley, M.L., Irving, K.E.: Connected classroom technology facilitates multiple components of formative assessment practice. J. Sci. Educ. Technol. **24**, 56–68 (2015)
110. Powell, C.B., Mason, D.S.: Effectiveness of podcasts delivered on mobile devices as a support for student learning during general chemistry laboratories. J. Sci. Educ. Technol. **22**, 148–170 (2013)
111. Schneps, M.H., Ruel, J., Sonnert, G., Dussault, M., Griffin, M., Sadler, P.M.: Conceptualizing astronomical scale: virtual simulations on handheld tablet computers reverse misconceptions. Comput. Educ. **70**, 269–280 (2014)
112. Sung, E., Mayer, R.E.: Online multimedia learning with mobile devices and desktop computers: An experimental test of Clark's methods-not-media hypothesis. Comput. Hum. Behav. **29**(3), 639–647 (2013)
113. Hsu, C.K., Hwang, G.J., Chang, Y.T., Chang, C.K.: Effects of video caption modes on English listening comprehension and vocabulary acquisition using handheld devices. Educ. Technol. Soc. **16**(1), 403–414 (2013)
114. Wong, L.H., Hsu, C.K., Sun, J., Boticki, I.: How flexible grouping affects the collaborative patterns in a mobile-assisted Chinese character learning game? Educ. Technol. Soc. **16**(2), 174–187 (2013)
115. Liu, M., Navarrete, C.C., Wivagg, J.: Potentials of mobile technology for K-12 education: an investigation of iPod touch use for English language learners in the United States. Educ. Technol. Soc. **17**(2), 115–126 (2014)
116. Kim, D., Kim, D.J.: Effect of screen size on multimedia vocabulary learning. Br. J. Educ. Technol. **43**(1), 62–70 (2012)
117. Huang, R.T., Jang, S.J., Machtmes, K., Deggs, D.: Investigating the roles of perceived playfulness, resistance to change and self-management of learning in mobile English learning outcome. Br. J. Educ. Technol. **43**(6), 1004–1015 (2012)
118. Viberg, O., Grönlund, A.: Cross-cultural analysis of users' attitudes toward the use of mobile devices in second and foreign language learning in higher education: A case from Sweden and China. Comput. Educ. **69**, 169–180 (2013)
119. de Jong, T., Specht, M., Koper, R.: A study of contextualised mobile information delivery for language learning. Educ. Technol. Soc. **13**(3), 110–125 (2010)
120. Şad, S.N., Göktaş, Ö.: Preservice teachers' perceptions about using mobile phones and laptops in education as mobile learning tools. Br. J. Educ. Technol. **45**(4), 606–618 (2014)
121. O'Bannon, B.W., Thomas, K.: Teacher perceptions of using mobile phones in the classroom: age matters! Comput. Educ. **74**, 15–25 (2014)

122. Abachi, H.R., Muhammad, G.: The impact of m-learning technology on students and educators. Comput. Hum. Behav. **30**, 491–496 (2014)
123. Mohammadi, H.: Social and individual antecedents of m-learning adoption in Iran. Comput. Hum. Behav. **49**, 191–207 (2015)
124. Ozdamli, F., Uzunboylu, H.: M-learning adequacy and perceptions of students and teachers in secondary schools. Br. J. Educ. Technol. **46**(1), 159–172 (2015)
125. Dündar, H., Akçayır, M.: Implementing tablet PCs in schools: students' attitudes and opinions. Comput. Hum. Behav. **32**, 40–46 (2014)
126. Teo, T., Noyes, J.: Development and validation of a computer attitude measure for young students (CAMYS). Comput. Hum. Behav. **24**, 2659–2667 (2008)
127. Kucirkova, N., Messer, D., Sheehy, K., Panadero, C.F.: Children's engagement with educational iPad apps: insights from a Spanish classroom. Comput. Educ. **71**, 175–184 (2014)
128. Bangert-Drowns, R.L., Pyke, C.: A taxonomy of student engagement with educational software: An exploration of literate thinking with electronic text. J. Educ. Comput. Res. **24**(3), 213–234 (2001)
129. Santos, P., Balestrini, M., Righi, V., Blat, J., Hernández-Leo, D.: Not interested in ICT? A case study to explore how a meaningful m-Learning activity fosters engagement among older users. In: Hernández-Leo, D., Ley, T., Klamma, R., Harrer, A. (eds.) EC-TEL 2013. LNCS 8095, pp. 328–342. Springer, Heidelberg (2013)
130. Hashim, K.F., Tan, F.B., Rashid, A.: Adult learners' intention to adopt mobile learning: a motivational perspective. Br. J. Educ. Technol. **46**(2), 381–390 (2015)
131. Stafford, T.F.: Understanding motivations for internet use in distance education. IEEE Trans. Educ. **48**(2), 301–307 (2005)
132. Norqvist, L., Jahnke, I., Olsson, A.: The learners' expressed values of learning in a media tablet learning culture. In: Rensing, C., de Freitas, S., Muñoz-Merino, P.J. (eds.) EC-TEL 2014. LNCS 8719, pp. 458–463. Springer, Switzerland (2014)
133. Haglund, B.: Stimulated recall: Några anteckningar om en metod att generera data. Pedagogisk forskning i Sverige **8**(3), 145–157 (2003)
134. Cappello, M.: Photo interviews: eliciting data through conversations with children. Field Methods **17**, 170–182 (2005)
135. Sevillano-García, M.L., Vázquez-Cano, E.: The impact of digital mobile devices in higher education. Educ. Technol. Soc. **18**(1), 106–118 (2015)
136. Tsai, P.S., Tsai, C.C., Hwang, G.H.: College students' conceptions of context-aware ubiquitous learning: a phenomenographic analysis. Internet High. Educ. **14**(3), 137–141 (2011)
137. Richardson, J.T.E.: The concepts and methods of phenomenographic research. Rev. Educ. Res. **69**(1), 53–82 (1999)
138. Tabuenca, B., Kalz, M., Ternier, S., Specht, M.: Stop and think: exploring mobile notifications to foster reflective practice on meta-learning. IEEE Trans. Learn. Technol. **8**(1), 124–135 (2015)
139. Muñoz-Organero, M., Muñoz-Merino, P.J., Delgado-Kloos, C.: Sending learning pills to mobile devices in class to enhance student performance and motivation in network services configuration courses. IEEE Trans. Educ. **55**(1), 83–87 (2012)
140. Ekanayake, S.Y., Wishart, J.: Integrating mobile phones into teaching and learning: a case study of teacher training through professional development workshops. Br. J. Educ. Technol. **46**(1), 173–189 (2015)
141. Sun, J.C.Y.: Influence of polling technologies on student engagement: an analysis of student motivation, academic performance, and brainwave data. Comput. Educ. **72**, 80–89 (2014)
142. Güler, Ç., Kılıç, E., Çavus, E.: A comparison of difficulties in instructional design processes: Mobile vs. desktop. Comput. Hum. Behav. **39**, 128–135 (2014)
143. Boot, E.W., van Merriënboer, J.J.G., Theunissen, N.C.M.: Improving the development of instructional software: three building-block solutions to interrelate design and production. Comput. Hum. Behav. **24**(3), 1275–1292 (2008)

144. Ryu, H., Parsons, D.: Risky business or sharing the load?—Social flow in collaborative mobile learning. Comput. Educ. **58**(2), 707–720 (2012)
145. Furió, D., González-Gancedo, S., Juan, M.C., Seguí, I., Rando, N.: Evaluation of learning outcomes using an educational iPhone game vs. traditional game. Comput. Educ. **64**, 1–23 (2013)
146. Lan, Y.F., Sie, Y.S.: Using RSS to support mobile learning based on media richness theory. Comput. Educ. **55**(2), 723–732 (2010)
147. Daft, R.L., Lengel, R.H.: Information richness: a new approach to managerial behavior and organization design. In: Staw, B., Cummings, L.L. (eds.) Research in Organizational Behavior, vol. 6, pp. 191–e233 (1984)
148. Kim, H., Lee, M., Kim, M.: Effects of mobile instant messaging on collaborative learning processes and outcomes: the case of South Korea. Educ. Technol. Soc. **17**(2), 31–42 (2014)
149. Fonseca, D., Martí, N., Redondo, E., Navarro, I., Sánchez, A.: Relationship between student profile, tool use, participation, and academic performance with the use of augmented reality technology for visualized architecture models. Comput. Hum. Behav. **31**, 434–445 (2014)
150. Souleles, N., Savva, S., Watters, H., Annesley, A., Bull, B.: A phenomenographic investigation on the use of iPads among undergraduate art and design students. Br. J. Educ. Technol. **46**(1), 131–141 (2015)
151. Martin, F., Ertzberger, J.: Here and now mobile learning: an experimental study on the use of mobile technology. Comput. Educ. **68**, 76–85 (2013)
152. Melero, J., Hernández-Leo, D., Manatunga, K.: Group-based mobile learning: do group size and sharing mobile devices matter? Comput. Hum. Behav. **44**, 377–385 (2015)
153. Gu, J., Churchill, D., Lu, J.: Mobile web 2.0 in the workplace: a case study of employees' informal learning. Br. J. Educ. Technol. **45**(6), 1049–1059 (2014)
154. Fuller, R., Joynes, V.: Should mobile learning be compulsory for preparing students for learning in the workplace? Br. J. Educ. Technol. **46**(1), 153–158 (2015)
155. Diaconita, I., Rensing, C., Tittel, S.: Getting the information you need, when you need it: a context-aware Q&A system for collaborative learning. In: Rensing, C., de Freitas, S., Muñoz-Merino, P.J. (eds.) EC-TEL 2014. LNCS 8719, pp. 410–415. Springer, Switzerland (2014)

Using Augmented Reality to Support Children's Situational Interest and Science Learning During Context-Sensitive Informal Mobile Learning

Heather Toomey Zimmerman, Susan M. Land and Yong Ju Jung

Abstract This research examines how augmented reality (AR) tools can be integrating into informal learning experiences in ways that support children's engagement in science in their communities. We conducted a series of video-based studies over 4 years in an arboretum and a nature center with families and children. In this study (the four iteration of the *Tree Investigators* design-based research project), 1-hour sessions were conducted at a summer camp for 6 weeks at Shaver's Creek Environmental Center. The sessions supported children to learn about the life cycle of trees with iPad computer tablets. Data collected included pre- and post-assessments and video records of children engaged in the science practice of observation. Analysis included the Wilcoxon signed-rank test of 42 paired assessments, the microethnographic analysis of transcripts of dyads and triads engaged with AR tools, and the creation of one case study of a pair of boys, who were representative of others in the dataset. Across the dataset, we found three sociotechnical interactions that contributed to triggering situational interests during the summer camp learning experience: (a) discoveries in the environment related to nature, (b) prior experiences that led to anticipation or expectation about what would happen, and (c) hands-on experiences with natural phenomenon. Implications of the study include that AR tools can trigger and maintain children's situational interest and science learning outcomes during context-sensitive informal mobile learning.

H.T. Zimmerman (✉) · S.M. Land · Y.J. Jung
Learning, Design and Technology Program, Penn State University,
317 Keller Building University Park, 16802 PA, USA
e-mail: heather@psu.edu

S.M. Land
e-mail: sland@psu.edu

Y.J. Jung
e-mail: yyj5102@psu.edu

© Springer International Publishing Switzerland 2016
A. Peña-Ayala (ed.), *Mobile, Ubiquitous, and Pervasive Learning*,
Advances in Intelligent Systems and Computing 406,
DOI 10.1007/978-3-319-26518-6_4

Keywords Augmented reality · Situational interest · Science learning · Technology-enhanced learning · Ubiquitous learning · Context-dependent learning · Context-sensitive learning · Informal learning · Digital photography · Environmental education

Abbreviations

AR   Augmented reality
ILI   Informal learning institution

1 Introduction

Modern perspectives on mobile learning recognize the situated nature of the sociocultural contexts in which learning occurs [1]. Context, in this sense, is broader than just location; context encompasses social, material, cultural, and environmental influences that shape and constrain what is possible to learn [2]. This perspective on context has influenced the design of technology-enhanced learning experiences for children in the informal education field.

Mobile technology can support context-sensitive learning by digitally augmenting learners' experiences with their surroundings to create new forms of learner–technology–setting interactions across time and space [3, 4]. Given the ubiquity and portability of mobile devices, they are well suited for creating learning spaces in everyday locations [1] such as nature centers, gardens, and community-based museums. Mobile devices also support the personalization of learning via on-demand resources and capitalize on learners' interests [5]. Learning about an area of interest can be distributed across settings, resources, people, technologies, and designed learning environments [5]. Advancing theory and design of context-sensitive informal mobile learning environments, then, entails intersecting sociotechnical considerations of the context. These sociotechnical considerations include learners' prior experiences, learners' interests, social interactions within the setting, the physical and environmental features of a setting, and available technologies [2, 3].

This chapter presents (a) theory related to context-sensitive learning, augmented reality (AR), and interest, (b) a review of empirical studies related to AR and ubiquitous computing in outdoor and community spaces, and (c) findings from our *Tree Investigators* research project that used AR tools and mobile computers in an informal learning setting to support learning about biology.

2   Theory Related to Context-Sensitive Learning, AR, and Supporting Children's Interest with Mobile Technologies

Learning is situated within specific social, material, and cultural contexts [6] with individual, social, and cultural components [7]. Informal learning happens through social interactions within museums, homes, and other everyday settings [8]. Context-sensitive, informal learning occurs in open learning environments [9, 10] where open learning environments include both the interactions that occur within one particular setting and the prior ideas and purposes of learners that occurred within another setting or context. In the open learning environment perspective, the learners' interpretations originate from their personal experiences. The designer using an open learning environment perspective emphasizes the mediating role of the individual in defining meanings and in establishing learning goals [9]. Open learning environments typically rely on technologies because their design requires the coordination of tools, resources, and pedagogies [10] to engage learners in the process of constructing knowledge.

2.1   Context-Sensitive Learning with Mobile Technologies

When used in context-sensitive informal learning, mobile technologies support learners' meaning making through providing continuous interaction between learners and setting [11, 12]. Mobile computers also can bridge different learning settings [13], thereby meeting the requirements of open learning environments. Previous studies related to context-sensitive mobile learning have identified two approaches to informal, open learning environments: (a) providing specific content at the right time and right place and (b) reinforcing learners' choices and experiences driven by their interests.

The first approach to mobile context-sensitive learning environments has focused on providing learning resources through mobile computers that can detect learners' locations. For example, Chen and colleagues [14] designed a context-sensitive ubiquitous learning system that delivered appropriate learning content based on a learner's location within a museum setting.

The second approach to mobile context-sensitive learning environments focuses on augmenting learning experiences by supporting learners' choices and interests. In our work, we take this second approach; we use an open learning environment perspective in an informal setting to allow children to select which aspect of a science topic to explore, based on their interests and experiences. Instead of prompting different learning content based on learners' locations, the AR mobile technology supports interest-driven interactions between learners and the nature center setting. This chapter focuses on this second approach to context-sensitive mobile learning, where designers augment learning experiences to support learners' choices and interests.

2.2 AR for Context-Sensitive Learning

AR is a technology that combines the physical setting with virtual information [15–17]. When designing AR with mobile technology for educational purposes, the connecting of digital augmentations to the natural setting in ways that support learning is critical. AR includes a broad range of tools. Early AR technologies used immersive devices such as helmets or goggles, but more recently, AR for learning has been conceptualized to include any technology that combines real and virtual experiences [12].

AR tools can include digital photographs, images, or text that direct the learners' attention to key details of the setting. AR tools can also be used to pose questions and prompts for conversation, enhancing the interactions on-site. Across the various design approaches that use AR tools to support context-sensitive learning, two main approaches are seen in the literature. The first approach to AR in education is to use AR tools to impose narratives, simulations, or games on a setting to engage the learners in educational activities. The second approach is to use AR tools to engage e learners with the disciplinary narratives within a setting.

AR Using Narratives, Simulations, or Games Imposed on a Setting. In AR games and simulations, learners interact with virtual objects or digital scenes that are linked to a physical setting [18]. For example, an interactive mystery game called *Mystery at the Museum* was designed for groups of parents and children at a museum [19]. With the AR game, learners solved crimes, which required exploring diverse artifacts in the museum. This game immersed learners in a virtual simulation, which helped them to engage the museum's exhibits more deeply as they collaborated with other learners.

Alien Contact! is an AR game that enhances learners' math, language arts, and scientific literacies by engaging learners in collaborative, participatory activities [20]. Middle and high school students play the game based on a fictional scenario of aliens landing on Earth, and learners are asked to discover the reasons for the aliens' arrival. Mobile computers and GPS software detect locations of learners and display corresponding digital objects and virtual people (i.e., avatars). In the *Alien Contact*! game, learners interview avatars to collect clues to discover the aliens' intentions. These activities afford learners to be immersed in the AR learning environment that is a combination of virtual elements and the real setting.

AR to Support Disciplinary Thinking within a Setting. In contrast to the approach of adding gaming narratives onto a physical setting, AR has also been used to support disciplinary thinking in a setting through incorporating scaffolding and just-in-time information. AR can support learners' engagement in science inquiry, discourse, and observations [15] through blending the physical setting with

Fig. 1 Two learners using *Tree Investigators* mobile AR app to engage in context-sensitive mobile computing during their summer camp experience

digital information. For example, *Musex* is a mobile AR app, which provides science-related questions to children related to the exhibits in a museum [21]. This project supported collaborative learning because pairs of learners participated in this activity together as well as communicated in real time with others.

AR tools also have been used to support engagement in science inquiry practices. For example, Looi et al. [22] developed mobile learning software called 3Rs to support young learners' science practices in everyday spaces, such as supermarkets. Throughout the sequence of activities (i.e., challenge, plan, experience, conclusion, and reflection), the technology tools prompted challenges that could be solved by observing scientific objects and then recording observations. Other projects have used mobile technologies to support similar inquiry processes such as ask, investigate, create, discuss, and reflect [23], tied to specific real-world locations.

Likewise, our research team has used mobile resources and AR to support learners' scientific observations in the outdoors, as shown in Fig. 1. Our focus is to use AR tools to support learners to make conceptual connections about scientific cycles they observe in their community, based on their observations of natural flora and fauna [3, 24].

2.3 Interest Development Within Context-Sensitive Learning

Because interest is one critical factor to drive learners to make choices to participate in informal educational activities, context-sensitive learning holds promise as a design theory for mobile learning experiences that support learners' interest development. Unlike developed long-term interests, situational interests are generated by specific characteristics of something that the learner observes or notices (i.e., a stimulus) [25]. In other words, situational interests are evoked by the environment surrounding learners [26, 27], which provides designers an opportunity to influence learners' situational interest. Thus, when designing context-sensitive mobile learning, we posit that AR can be added to mobile computer apps to support the emergence of situational interests through supporting learners to observe and notice something that will capture their curiosity.

According to the four-phase model of interest development [27], interests develop through stages over time—as long as learners maintain their involvement in the activities related to their interest. First, learners' situational interests are triggered by environmental stimuli. Second, learners' interests are maintained through meaningful involvement in activities. Third, maintained situational interests can emerge into individuals' interests, and then in the final stage, people have created well developed and stable long-term interests.

The first phase of the Hidi and Renninger model [27] has unique potential for AR technology to support children's interests. AR tools can enhance the triggering of children's situational interests, by focusing the children's attention to environmental and social elements that they may find intriguing or curious. In addition, AR tools can highlight a setting's environmental features with incongruous or surprising information, which may also trigger the situational interests of learners. Finally, AR technologies can facilitate positive emotional reactions toward a particular situation as well as support sustained cognitive engagement, which are both seen as the evidence of situational interests [27, 28].

Studies of informal science learning have examined the role of out-of-school settings in interest development. Zimmerman and Bell [29] used cross-setting ethnographic fieldwork to find that children were interested in participating in scientific practices (including observing) when these practices were related to children's interests found in their everyday activities. Other research confirms that multiple informal settings can trigger and support interest related to science including in homes [30], zoos and aquaria [28], and afterschool clubs [31]. Triggering and maintaining interests has been shown to be important in relation to academic achievement, identification toward science, and eventual participation in STEM careers [32–34]. In this regard, our work importantly brings together informal science learning with context-sensitive AR to enhance outdoor education in ways that situational interests can be triggered and maintained along with science knowledge and practices.

3 Tree Investigators: AR and Ubiquitous Computing Supporting Science Learning in an Outdoor, Community Setting

Building on theory from educational technology, informal learning, and situational interest, we developed design considerations to guide the development of *Tree Investigators* as a context-sensitive informal mobile learning experience. We first describe these design considerations and how they guided our design decisions.

Second, we describe the *Tree Investigators* project to show how these design considerations were manifest in the resulting AR mobile technology.

3.1 Design Considerations for Using AR for Learning in Outdoor Community Spaces

Our primary perspective is that designing mobile AR technology to support context-sensitive learning in an outdoor community space requires a localized perspective on design. A localized perspective on design can be referred to as place-based education [35]. Place-based education engages children and families in learning activities that have strong community connections [36]. Researchers and educators adopt place-based education to highlight the disciplinary information within a place, moving from abstracted knowledge to relevant, local knowledge [37] that is related to communities' histories [38]. Place-based education can be used in context-sensitive informal mobile learning so that learners connect new scientific ideas to their community-based experiences outdoors. We adopt the view that place-based learning in ILIs can connect out-of-school learners to their communities via mobile computers, as shown in Fig. 2.

Second, we consider possible technical barriers that can happen particularly in outdoor spaces when using mobile computers. For example, Internet connection is often not stable in forest settings or parks that are outside of urban centers. Consequently, designers working with parks and nature centers may need to adopt mobile technology that enables learners to access learning materials without the Internet. Given the importance of context-sensitive learning and community settings, *Tree Investigators* offers a design model that can support the use of mobile, ubiquitous technologies in parks, state forestlands, and nature center settings.

Fig. 2 The *Tree Investigators* app showing place-based images connecting learners in summer camp to their local community's trees

Fig. 3 The learners used AR to learn about a tree's life cycle, such as the seed stage

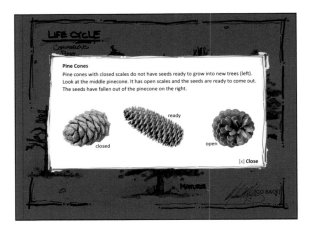

Third, AR activities should be designed to support learners' interactions with environmental settings, in support of disciplinary practices such as observation. For example, at museums or nature centers, learners may have difficulty distinguishing what is scientifically relevant to observe among the numerous objects; therefore, AR can support learners to notice relevant disciplinary phenomena [41]. When supporting disciplinary practices within informal settings, designers should avoid fostering "heads-down" engagement [39, 40] with screens, where learners engage with devices rather than the setting around them. In our work, we used digital photography, such as Fig. 3, to accomplish this observational support so that learners remained engaged with the outdoors on the nature trail setting.

Fourth, AR can be designed to contain some elements that can evoke situational interests. For example, strategies for stimulating situational interests [28, 42], include using original materials; channeling learners' attention to surprising, curious, or novel phenomenon; using a pedagogy with hands-on activities; and developing materials that provide individual choice and engaging social interactions. Importantly, in keeping with our third design consideration, the AR tools used for simulating situational interests should focus the learner to attend to the unique aspects of the informal setting, rather than only the device's screen.

3.2   The Technological Innovation of Tree Investigators Project

Our research team developed *Tree Investigators* as a design-based research project for tablet-mediated collaborative science learning in outdoor learning environments [3, 17]. Our first iteration focused on supporting engagement in science practices [24]. We later added perspectives of digital making to support science knowledge generation in iterations two and three [3]. In the fourth iteration of the project that

Fig. 4 Two children use the *Tree Investigators* iPad app's AR viewfinder and checklist to support their observation of trees

we describe here, we implemented AR technology to support children to identify trees at various stages of their life cycles at a nature center summer camp. One goal of iteration four was to apply design considerations for supporting situational interest and engagement in science practices via AR tools.

In iteration four, we developed an AR-enhanced mobile app with resources optimized for iPads that allowed for real-time viewing of the physical environment with overlays of digital photographs, text, conceptual models, and supports for scientific observation practices. Science content was organized by the use of a graphic organizer of tree life cycles where learners touched part of an image to learn more. Images and text directed learners look around them at the natural setting for evidence of their claims, supporting both multisensory and multimodal context-sensitive learning experiences. Led by a naturalist, groups of learners initially explored the nature center as a large group. Then, in small groups of dyads or triads, they used the AR tool we developed to identify evidence of tree life cycles (Fig. 4).

The AR tool supported observation through superimposing criteria (in the form of check boxes) for identifying trees at all stages of their life cycle. These criteria were viewable as AR supports within the camera viewfinder. Thus, when learners viewed a tree specimen through the AR viewfinder and took a photo of this tree, the photo saved the criteria that learners checked on the photograph.

Learners then compiled their photographs into a personalized tree life cycle digital artifact while they were at the nature center, as shown in Fig. 5. The criteria the learners applied through the AR viewfinder were also visible in the digital artifact (see Fig. 5). Through engaging in digital making of a personalized tree life cycle, children enacted the place-based [35–38] goals of the project by turning abstracted tree life cycle knowledge into local knowledge in their community.

Fig. 5 A tree life cycle photo collage

4 Project Findings Related to the Connection Between a Context-Sensitive AR Approach and Learners' Interests

The overarching research methodology driving the *Tree Investigators* project is design-based research [43–45], which informs theory and practice through iterative implementations of an intervention in a real-world educational setting. For the past 5 years, our team worked in partnership with Shaver's Creek Environmental Center and the Arboretum at Penn State on educational programming for children and families [3, 24]. Across our research design, our goal is to design mobile technologies to augment scientifically meaningful experiences for out-of-school time in outdoor informal learning institutions (ILIs) [8, 46].

4.1 Methodology

Our fourth design-based research iteration was based on data collected from seven, 1-h sessions during a nature summer camp over two weeks at Shaver's Creek Environmental Center. As described in the previous section, the sessions were designed to support children to learn about tree life cycles with AR and iPad mini tablets. The mobile app included a conceptual organizer to learn about the life cycle of trees and AR tools for capturing evidence of tree life cycles and making a digital collage with the pictures taken by learners. The sessions included three phases of learning activities: (a) a naturalist's structured exploration of tree life cycles with a mobile app and preselected trees; (b) learner-centered exploration and identification of trees for each stage of the life cycle, and (c) digital artifact-making in small groups with the photographs collect during the camp.

Setting and Participants. The participants were 42 children (18 females, 24 males), who signed up for a weeklong nature summer camp as part of normal summer leisure activities and to meet parent's childcare needs. The children were nine to 12-years old (which typically corresponds to entering fourth through six grades in the United States). All 42 children consented to participate in the activity and the pre- and post-test. A subset of the children ($n = 35$) provided their assent to be video recorded during the 1-h program.

Data Collection. Data were collected during normal camp activities. Children first answered a five-question assessment about tree life cycle concepts, which lasted approximately 4 min. Children then were video recorded as they participated in an informal education program that lasted approximately 45–50 min. Children next participated in a short interview with the researchers about their experience. Finally, learners completed the same tree life cycle concept assessment as a post-test. Although collaborative team activities comprised the main focus of the camp workshop, the pre- and post-tests were administered and scored individually.

Data Analysis. To answer the research question, *How can learners' knowledge of tree life cycle concepts be supported along with their situational interests related to nature via an AR mobile app?*, we employed multiple analytical tools. As noted previously, our first analytical strategy was to administer and score the tree life cycle concept assessment. Five open-ended questions were used identically for the pre- and post-test. The questions asked about facts regarding the life cycle for trees (e.g., 'List the stages of the life cycle for trees') or characteristics for some of the stages of trees (e.g., "What are some of the differences between sapling and mature trees?"). Learners' answers were assessed out of a total possible score of 19. Because we focused on their knowledge of the life cycle, simple misspelled words were scored as correct (i.e., "moture" tree was scored the same as was the correct spelling of "mature" tree).

Our second analytical technique was a microethnographic [47] line-by-line investigation of video data to investigate how and when situational interests were triggered with support of AR in the outdoor nature center. To identify how and when situational interests were triggered in the dataset, theory guided our coding. Episodes were strategically sampled when the participants' emotional exclamation, feeling of enjoyment, focused attention and engagement, and committed talk [28] were identified. This chapter focuses on the case of one group of two boys who participated in the camp, Jacob and Pavel. These two boys were strategically sampled because their behaviors were typical of those of others in the dataset. They were chosen from other typical cases because their talk was more suitable for excerpting segments for publication because their talk was less interrupted by other children, included more direct references to the site in ways that made their talk transparent without the video, and their talk included humorous elements.

4.2 Findings from the Tree Investigators Mobile AR Outdoor Education Program: Knowledge Gains About Trees Across 42 Summer Camp Participants

To answer the first part of our research question related to science learning, the 42 participating children were given a five-item assessment of tree life cycle knowledge and concepts, before and after the *Tree Investigators* intervention. The mean score for pretest was 5.07 (SD = 2.91) out of 19. The mean score for post-test was 11.14 (SD = 2.94) out of 19. The learners, on average, increased their understanding of the tree life cycle by 6.5 points (see Table 1).

Due to the small number of participants drawn from a specific summer camp, we could not assume a normal distribution of participants, requiring nonparametric statistics. Hence, the Wilcoxon signed-rank test was conducted in lieu of a paired *t*-test. The Wilcoxon signed-rank test results show that the post-test scores are significantly higher (Mdn = 12.00) than the pretest scores (Mdn = 5.50), $z = -5.64$, $p < 0.001$, $r = -0.62$, suggesting improvements in the children's understandings of concepts about the tree life cycle (see Fig. 6). Cohen's $d = 2.22$ and the effect-size correlation was $r = 0.743$, suggesting that a 6.5 increase demonstrates a large effect size.

Table 1 The results of the 5-item (19 points) assessment (n = 42 children) of tree life cycle concepts

	N	Median	Mean	Std. deviation	Min	Max
Pre-test	42	5.50	5.07	2.91	0	11
Post-test	42	12.00	11.14	2.94	1	16

Fig. 6 The differences between the pre- and post-test were significant statistically, with a large effect size

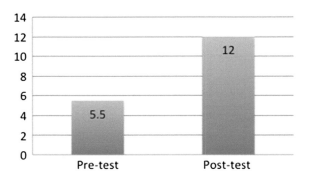

4.3 Qualitative Case Study Findings: Triggering Situational Interests with AR in an Outdoor Space

To answer the second part of our research question related to situational interest, we examined all episodes that were coded based on prior studies of interest [27, 28]. We examined episodes that included emotional exclamations, focused attention, and committed talk. Across the dataset, three situations emerged as patterns that contributed to trigger situational interests during the learning activities: (a) discoveries in the environment related to nature that were sparked by *Tree Investigators* AR tools, (b) prior experiences that led to anticipation or expectations about what would happen, and (c) hands-on experiences with natural phenomenon that were generated by the prompts given in *Tree Investigators* AR tools. To illustrate these three findings, we present a case study of Jacob and Pavel as they worked with the *Tree Investigators* AR mobile app.

Discoveries in the Environment Supported by the AR Tool Triggered Interest. Across the dataset, situational interests were triggered when the children discovered something on the trails that was related to the disciplinary content of the *Tree Investigators* AR and mobile materials, such as descriptions and photographs of specific trees. Most children expressed their excitement verbally or with gestures and physical movements such as pointing or jumping up and down. Because the learning activity supported by the *Tree Investigators* AR tools required the participants to find specific examples of each stage of the tree life cycle, interest-related events connected to tree life cycle discoveries were the most common. Jacob and Pavel illustrate this finding in two separate episodes below, when the two boys discovered a pinecone and a seedling.

> *Jacob and Pavel's interest was triggered when they found a pinecone*:
> Jacob: ((reads the text from the mobile app)) "Seed or fruit or berry or cone or nut..."
> Pavel: Oh, berries are easy! ((keeps looking around with Jacob)) Where is it? I am looking for a pinecone. Somebody give me a pinecone. Gimme gimme a pinecone.
> Pavel: Pinecone! ((grabs one from the ground))
> Jacob: Oh good!
> ((Pavel picks up the pinecone))
>
> *Jacob and Pavel's interest was triggered when they found a seedling*:
> Pavel: Okay so we got the seeds, right?
> Jacob: Yeah ((looking down the mobile app)) now we have to get a seedling.
> Pavel: Seedling! Well, right over here! ((gesturing digging in dirt)) Neeerrnerr nee nee nee!

The boys' interest in pinecones was initiated when they read the text from the mobile app about the types of seeds and seed containers on trees (e.g., fruits, cones). After rejecting berries, Pavel began a search for a pinecone ("gimme a pinecone"). The mobile app supported them to explore the nature center's trails and find the scientific phenomenon that the children were interested in identifying through providing digital photographs of possible specimens to find. In this way, the mobile app reinforced the interaction between the local setting (i.e., pinecones on the nature trail) and the learners. When they found the appropriate species throughout the *Tree Investigators* learning experiences, their situational interests were explicitly expressed with positive exclamation toward the pinecones.

Similarly in the second seedling example, the mobile app guided the boys to find something that was both scientifically relevant and interesting to them—a seedling. And when they found the seedling after searching and observing, their situational interests were evoked with the verbal and gestural expression of exclamation and pleasure ("Seedling! Well, right over here!" and digging gesture) as well as a sense of satisfaction from completing the task ("Neeerrnerr nee nee nee!").

Anticipation and Expectations about Nature-Triggered Interest We also found evidence in our dataset that the learners expressed interest about a natural object based on the prior knowledge that they anticipated or expected to find, and then they set out to find the anticipated object. When this occurred, the learners expressed interests related to seeing on the nature trails based on what they knew related to their participation in the context-sensitive informal mobile learning activity.

To illustrate this finding, we return to the first excerpt, when Pavel and Jacob set out to find a type of seed. Pavel showed confidence in his ability based on his prior experiences with certain seed types ("Oh, berries are easy!"), and then participated with his partner in finding a pinecone with eager excitement. This excerpt showed the anticipation that the learners developed from their prior experiences. Pavel and Jacob expected that they could easily perform the activity, which triggered situational interests in the activity—especially to complete the activity quickly ("Gimme gimme a pinecone"). This episode ended with picking up the pinecone and in many cases in our dataset, the learners touched or picked up the object of interest as we discuss below.

Hands-on, Sensory Experiences with Natural Phenomenon that were Supported by AR-Tools Triggered Learners' Interest. The intention of our *Tree Investigators* design was to connect interactions between the natural setting and learners, in ways that enhanced the children's situational interest and science learning. In some cases, learners' science learning was supported by the app; but, hands-on experiences with the trees, which the app discussed, were often the triggers for situational interest. The following excerpt is an exemplar of this finding as Pavel's situational interest was triggered when he touched leaves and experienced the leaves' texture.

Pavel became interested in leaves' texture:
Pavel: Can I touch these [leaves]? Jacob, come feel this!
Jacob: Yeah. ((Walks to Pavel but looking at the content on the tablet)) Hey, this is the same. That doesn't…that doesn't make any sense!
Pavel: Jacob! Feel this! ((Gestures to leaves))
Jacob: Oh. ((Maintains focus on tablet)) I will just take picture of it…
Pavel: So soft!

In this example, Pavel's interest was triggered about this tree based on the texture of the leaf ("So soft!"), as evidenced by his emotional exclamation. Because his situational interest was evoked, he sustained lengthy engagement at the tree and tried to engage his peer Jacob in the same activity. Given that AR and app material encouraged visual and tactile observation (i.e., learners tested if they could put their hands around a tree trunk to check its girth to estimate its age), many learners began to explore the trees around them with multiple senses. Most children in our dataset made tactile observations as well as visual observations.

5 Implications for Context-Sensitive Informal Mobile Learning to Foster Situational Interest and Science Learning

This chapter illustrated how ubiquitous mobile technology supported context-sensitive science learning and the early stages of interest development of learners. Within the *Tree Investigators* project, we found that learners showed an increased understanding of the tree life cycle as illustrated by significant gains in pre- and post-test scores. Also, the video-based microethnographic analysis showed that learners' situational interests were triggered and supported within the 1-h informal mobile learning program during summer camp. In our analysis, we found that the AR checklist and the photographic artifacts were the two technological tools within our app that helped learners focus their observations in ways that were consistent with the four design considerations for context-sensitive mobile informal learning of outlined in Sect. 4.1. We discuss these two tools below, and then we end with a discussion of other sociotechnical interactions that triggered the children's situational interests.

5.1 The AR Checklist Supported Learning and Interest Development

The checklist served for most groups of learners as a just-in-time, in-the-moment reminder of the key criteria for distinguishing different stages of a tree's life cycle. In peer interactions, the checklist served various roles to support science learning and interest development. The checklist encouraged children to test criteria shown on the checklist on actual trees—encouraging sensory engagement with nature and social engagement with each other, as shown in Sect. 5.3 when Jacob and Pavel focused on the pinecone.

We found these interactions as evidence that our "heads-up" interaction goals with the site were achieved [39, 40]. The criteria on the checklist allowed the children to check and double-check each others' understandings and have an authority to fall back on to develop final categorization of the tree life cycle stages. It also support the young people to look close at trees, creating curiosities and wonderings that we found to be aligned with the triggering of situational interests.

5.2 Photography and the Digital Making of Photo Artifacts

Digital photography functionality was most relevant to support science learning as shown through the science talk and observational practices of the children. Taking digital photos was a shared task in *Tree Investigators*, and taking the photos was a highly enjoyable activity for all groups. The photography was closely tied to our goal of supporting children to learn to see like a scientist through supporting the youth to discern the different stages of the life cycle. Importantly, the photographs served as a shared knowledge artifact around which learners oriented their conversations and observational practices.

In order to take pictures, the children had to describe the location and type of tree that they saw. This observationally focused talk was eventually used as evidence (by most groups) to support descriptions and emerging claims about a tree's life cycle stages. While our prior work has demonstrated it is useful to connect parents to children's summer camp experiences in order to support learning [17], the findings of *Tree Investigators* iteration four here show that it is important to support peer interactions around the ecology content in the outdoors as well.

Near the end of the summer camp program, the learners worked in teams to engage in a short digital making activity on the iPads. The learners' task was to use multiple digital photographs their team had taken on the trails to make a customized and personalized tree life cycle image. Through engaging in digital making of a tree life cycle, we found that the children connected what they experienced in their local nature center related to the themes of the discipline of biology.

5.3 Sociotechnical Interactions that Contributed to Trigger Situational Interests

Our analysis revealed three sociotechnical interactions that contributed to trigger situational interests during the learning activities with the mobile technology: (a) discoveries in the environment related to nature were supported by the AR tool, (b) prior experiences led to anticipation or expectations about what would happen or what they would find, and (c) hands-on experiences with natural phenomenon were prompted by the AR tool and other app content. Taken together, these three findings suggest that our approach to context-sensitive ubiquitous learning, which included prompting location-specific science content and supporting learner-driven experiences within the setting, can trigger situational interests and support science learning.

Derived from these three findings, we suggest three strategies for designers and researchers seeking similar approaches to foster situational interest and informal science learning. First, given that the AR tool supported the learners' discoveries in the environment related to nature, we suggest that AR can be used for more than gamification and simulations—learners' interests in and observations of the natural world can be triggered and supported via AR.

Second, we found that learners had prior experiences in the outdoors that led to anticipation or expectations about what they would find or what would happen. Our results suggest an iterative process to design can help developers create prompts and digital materials that elicit prior knowledge in order to bridge the learners' past experiences to the new informal science learning experience.

Finally, given that the AR tool prompted sensory exploration and hands-on experiences, we argue that multisensory engagement in the outdoors can be supported by ubiquitous technologies [48]. Children's engagement in nature and with mobile computers can work together to support outdoor exploration and enjoyment. This last strategy of using AR to support tactile and visual observation of the natural setting is apt given recent calls for children to spend more time outdoors [49]. In short, our context-sensitive informal mobile learning technology supported the children to engage in multimodal sensory observations during summer camp, with time spent away from the iPad screen. We found that *Tree Investigators* allowed learners to engage with the natural world around them.

Acknowledgements We would like to thank the children who participated in our study, Shaver's Creek Environmental Center, and our funder Penn State's Center for Online Innovation in Learning (COIL). We acknowledge the contributions of our team members (http://sites.psu.edu/augmentedlearning/) who participated in the current and prior iterations of *Tree Investigators*: Brian J. Seely, Michael R. Mohney, Jaclyn Dudek, Jessica Briskin, Chrystal Maggiore, Soo Hyeon Kim, Gi Woong Choi, Fariha H. Salman, and Lucy R. McClain.

References

1. Sharples, M., Pea, R.D.: Mobile Learning. In: Sawyer, R.K. (ed.) Cambridge handbook of the learning sciences (2nd edition), pp. 1513–1573. Cambridge University Press, New York, NY (2014)

2. Pea, R.D.: Practices of distributed intelligence and designs for education. In: Salomon G. (ed.) Distributed cognitions (pp. 47–87). Cambridge University Press, New York, NY (1993)

3. Land, S.M., Zimmerman, H.T.: Socio-technical dimensions of an outdoor mobile learning environment: a three-phase design-based research investigation. Educ. Tech. Res. Dev. **63**(2), 229–255 (2015)

4. Sharples, M.: Forward to education in the wild. In: Brown E. (ed.) Education in the wild: contextual and location-based mobile learning in action (2010)

5. Barron, B.: Interest and self-sustained learning as catalysts of development: a learning ecology perspective. Hum. Dev. **49**(4), 193–224 (2006)

6. Brown, J.S., Collins, A., Duguid, P.: Situated cognition and the culture of learning. Educ. Researcher **18**(1), 32–42 (1989)

7. Rogoff, B.: The cultural nature of human development. Oxford University Press, London, UK (2003)

8. Falk, J.H., Dierking, L.D.: Learning from museums: visitor experiences and the making of meaning. AltaMira Press, Walnut Creek, CA (2000)

 9. Hannafin, M.J., Land, S.M., Oliver, K.: Open learning environments: foundations and models. In: Reigeluth, C. (ed.) Instructional design theories and models, vol. 2, pp. 115–140. Erlbaum, Mahway, NJ (1999)
10. Hannafin, M.J., Hill, J.R., Land, S.M., Lee, E.: Student-centered, open learning environments: research, theory, and practice. In: Spector, M., Merrill, M.D., Merrienboer, J., Driscoll, M. (eds.) Handbook of research for educational communications and technology, pp. 641–651. Routledge, London, UK (2013)
11. Cahill, C., Kuhn, A., Schmoll, S., Lo, W.T., McNally, B., Quintana, C.: Mobile learning in museums: how mobile supports for learning influence student behavior. In: Proceedings of the International Conference on Interaction Design and Children. 10, 21–28, USA (2011)
12. Klopfer, E.: The importance of reality. In Augmented learning: research and design of mobile educational games. The MIT Press, London, UK (2008)
13. Walker, K.: Designing for museum learning: visitor-constructed using mobile technologies. In: Luckin, R., Puntambekar, S., Goodyear, P., Grabowski, B., Underwood, J., Winters, N. (eds.) Handbook of design in educational technology, pp. 322–335. Routledge, New York, NY (2013)
14. Chen, C.C., Huang, T.C.: Learning in a u-museum: developing a context-aware ubiquitous learning environment. Comput. Educ. 59(3), 873–883 (2012)
15. Dunleavy, M., Dede, C.: Augmented reality teaching and learning. In: Spector, J.M., Merrill, M.D., Elen, J., Bishop, M.J. (eds.) Handbook of research on educational communications and technology, pp. 735–745. Springer, New York, NY (2014)
16. Milgram, P., Kishino, F.: A taxonomy of mixed reality visual displays. IEICE Trans. Inf. Syst. E77-D, 1–15 (1994)
17. Zimmerman, H.T., Land, S.M., Mohney, M., Choi, G., Maggiore, C., Kim, S., Jung, Y.J., Dudek, J.: Using augmented reality to support observations about trees during summer camp. In: Proceedings of the 14th International Conference on Interaction Design and Children, pp. 395–398. ACM, New York, NY (2015)
18. Yuen, S., Yaoyuneyong, G., Johnson, E.: Augmented reality: an overview and five directions for AR in education. J. Educ. Technol. Dev. Exch. 4(1), 119–140 (2011)
19. Klopfer, E., Perry, J., Squire, K., Jan, M.-F., Steinkuehler, C.: Mystery at the museum: a collaborative game for museum education. In: Proceedings of the International Conference on Computer Supported Collaborative Learning, pp. 316–320 (2005)
20. Dunleavy, M., Dede, C., Mitchell, R.: Affordances and limitations of immersive participatory augmented reality simulations for teaching and learning. J. Sci. Educ. Technol. 18(1), 7–22 (2009)
21. Yatani, K., Onuma, M., Sugimoto, M., Kusunoki, F.: Musex: a system for supporting children's collaborative learning in a museum with PDAs. Syst. Comput. Jpn. 35(14), 54–63 (2004)
22. Looi, C.-K., Seow, P., Zhang, B., So, H.-J., Chen, W., Wong, L.-H.: Leveraging mobile technology for sustainable seamless learning: a research agenda. Br. J. Educ. Technol. 41(2), 154–169 (2010)
23. Chiang, T.H., Yang, S.J., Hwang, G.J.: Students' online interactive patterns in augmented reality-based inquiry activities. Comput. Educ. 78, 97–108 (2014)
24. Zimmerman, H.T., Land, S.M., McClain, L.R., Mohney, M.R., Choi, G.W., Salman, F.H.: Tree investigators: supporting families and youth to coordinate observations with scientific knowledge. Int. J. Sci. Educ. 5(1), 44–67 (2015)
25. Krapp, A., Hidi, S., Renninger, K.A.: Chapter 1 Interest, learning, and development. In: Renninger, K.A., Hidi, S., Krapp, A. (eds.) The role of interest in learning and development, pp. 3–25. Lawrence Erlbaum Associates, NJ (1992)
26. Hidi, S.: Interest and its contribution as a mental resource for learning. Rev. Educ. Res. 60(4), 549–571 (1990)
27. Hidi, S., Renninger, K.A.: The four-phase model of interest development. Educational Psychologist. 41(2), 111–127 (2006)

28. Dohn, N.B.: Situational interest of high school students who visit an aquarium. Sci. Educ. **95**(2), 337–357 (2011)
29. Zimmerman, H.T., Bell, P.: Where young people see science: everyday activities connected to science. Int. J. Sci. Educ. **4**(1), 25–53 (2014)
30. Zimmerman, H.T.: Participating in science at home: recognition work and learning in biology. J. Res. Sci. Teach. **49**(5), 597–630 (2012)
31. Basu, S.J., Calabrese Barton, A.: Developing a sustained interest in science among urban minority youth. J. Res. Sci. Teach. **44**(3), 466–489 (2007)
32. Dabney, K.P., Tai, R.H., Almarode, J.T., Miller-Friedmann, J.L., Sonnert, G., Sadler, P.M., Hazari, Z.: Out-of-school time science activities and their association with career interest in STEM. Int. J. Sci. Educ. **2**(1), 63–79 (2012)
33. Leibham, M.B., Alexander, J.M., Johnson, K.E.: Science interests in preschool boys and girls: relations to later self-concept and science achievement. Sci. Educ. **97**, 574–593 (2013)
34. Tai, R.H., Lui, C.Q., Maltese, A.V., Fan, X.: Planning early for careers in science. Science **312**, 1143–1144 (2006)
35. Sobel, D.: Place-based education: connecting classroom and community. The Orion Society, Great Barrington, MA (2004)
36. Smith, G.A.: Place-based education learning to be where we are. Phi Delta Kappan **83**(8), 584–594 (2002)
37. Lim, M., Barton, A.C.: Science learning and a sense of place in a urban middle school. Cult. Sci. Educ. **1**, 107–142 (2006)
38. Gruenewald, D.A.: The best of both worlds: a critical pedagogy of place. Educ. Researcher **32**(4), 3–12 (2003)
39. Hsi, S.: A study of user experiences mediated by nomadic web content in a museum. J. Comput. Assist. Learn. **19**(3), 308–319 (2003)
40. Lyons, L.: Designing opportunistic user interfaces to support a collaborative museum exhibit. In: Proceedings of the 9th International Conference on Computer Supported Collaborative Learning, **1**, pp. 375–384. International Society of the Learning Sciences, (2009)
41. Zimmerman, H.T., Land, S.M.: Facilitating place-based learning in outdoor informal environments with mobile computers. Tech. Trends. **58**(1), 77–83 (2014)
42. Holstermann, N., Grube, D., Bögeholz, S.: Hands-on activities and their influence on students' interest. Res. Sci. Educ. **40**(5), 743–757 (2009)
43. Barab, S., Squire, K.: Design-based research: Putting a stake in the ground. J. Learn. Sci. **13**(1), 1–14 (2004)
44. Sandoval, W.A., Bell, P.: Design-based research methods for studying learning in context: introduction. Educ. Psychol. **39**(4), 199–201 (2004)
45. Hoadley, C.M.: Methodological alignment in design-based research. Educ. psychol. **39**(4), 203–212 (2004)
46. Bell, P., Lewenstein, B., Shouse, A.W., Feder, M.A., (eds.).: Learning science in informal environments: people, places, and pursuits. National Academies Press. (2009)
47. Spradley, J.P.: Participant observation. Holt, Rinehart, and Winston, New York, NY (1980)
48. Land, S.M., Zimmerman, H.T., Choi, G.W., Seely, B.J., Mohney, M.R.: Design of mobile learning for outdoor environments. In: Orey, M., Branch, R. (eds.) Educational media and technology year book, pp. 101–113. Springer, Heidelberg, Germany (2015)
49. Louv, R.: Last child in the woods: saving our children from nature-deficit disorder. Algonquin Books, Chapel Hill, NC (2008)

Cooperative Face-to-Face Learning with Connected Mobile Devices: The Future of Classroom Learning?

Martin Ebner, Sandra Schön, Hanan Khalil and Barbara Zuliani

Abstract Communication and collaboration among peers influence learning outcomes in a positive way. Therefore, our research work focuses on enhancing face-to-face group learning with the usage of mobile devices by developing a learning game for iPhone/iPad devices called MatheBingo. The app allows up to four learners to connect to each other through their mobile devices and learn together in a face-to-face setting. An initial evaluation in this field of research indicates the usefulness of such activities and how they uniquely motivate children to learn. It can be summarized that the connection of mobile devices is an important step toward the future of face-to-face classroom learning.

Keywords Collaboration · Math · Mobile learning · iOS development · Field study

Abbreviation list

ADS	Apple Definition Statement
HCI	Human computer interface
HY	Hypothesis
IICM	Institute for Information Systems and Computer Media
iOS	internet Operating System

M. Ebner (✉)
Educational Technology, Graz University of Technology, Graz, Austria
e-mail: martin.ebner@tugraz.at

S. Schön
Salzburg Research, Salzburg, Austria
e-mail: sandra.schoen@salzburgresearch.at

H. Khalil
Manosoura University, Mansoura, Egypt
e-mail: drhanan.khalil@gmail.com

B. Zuliani
Elementary School Breitenlee, Vienna, Austria
e-mail: barbara.zuliani@icloud.com

© Springer International Publishing Switzerland 2016
A. Peña-Ayala (ed.), *Mobile, Ubiquitous, and Pervasive Learning*,
Advances in Intelligent Systems and Computing 406,
DOI 10.1007/978-3-319-26518-6_5

121

Mathe Bingo	Math Bingo
mLearning	Mobile learning
TEL	Technology Enhanced Learning
TU Graz	Graz University of Technology
Wi-Fi	Wireless Fidelity

1 Introduction

The field of Technology Enhanced Learning (TEL) is a rapid emerging one. Just 10 years ago, TEL researchers introduced Web 2.0 [1] and called it e-Learning 2.0 [2, 3]. Since then Wikis [4, 5], Weblogs [6, 7], and Podcasts [8, 9] have become an essential part of today's classrooms. With the invention of smartphones mobile learning comes more and more attractive and nowadays we are thinking even about the usage of wearable devices for teaching and learning [10].

Nevertheless, with the rapid development of network communication technologies there has been an increase in the quantity of research on applying mobile technologies to learning. In terms of educational application, mobile technologies can be regarded as services that electronically deliver general and educational content to learners regardless of location and time [11]. Researchers have argued that mobile technologies have created many new and exciting opportunities for learning [12]. They provide instant learning feedback and guidance and use new interfaces for diverse learning approaches [13].

Within this chapter we present a new learning game for iPhone and iPads called "Mathe Bingo" ("Math Bingo") using an innovative collaborative learning setting for young learners (6–7 years of age) within classrooms. Additionally, we show and reflect upon results of a first trial within an Austrian so-called "iPad class" from both the observers' and the children's perspectives. First of all, we will start with the theoretical background and experiences with cooperative learning with mobile devices and the experiences with iPhone/iPad app development at Graz University of Technology (TU Graz), especially concerning apps for young learners.

Our chapter is organized as following: First of all, we describe the theoretical background and give insight to current studies about cooperative learning with a special eye on mobile devices. Afterwards the goal of our study, the research questions as well as the research method are presented. Section 5 points out the concept and design of our app-prototype. In addition we tested the app for the first time in classroom and carried our experiences. Finally, the discussion section describes our findings and the conclusion summarizes the research work and gives an outlook for future work.

2   Theoretical Background and Studies on Cooperative Learning with Mobile Devices in Classrooms

In this section we like to give a short introduction to cooperative and collaborative learning, how mobile devices can be used in classroom, and how games can facilitate cooperative and mobile learning. Finally, we point out our first experiences with iPhone development at TU Graz.

2.1   Cooperative Peer Learning in the Classroom

Cooperative and collaborative peer learning has been frequently seen as a stimulus for cognitive development, through its capacity to stimulate social interaction and learning among the members of a group. The goal of collaborative learning is to facilitate teaching through a coordinated and shared activity, by means of social interactions among the group members [14]. These social interactions are essential to achieve the desired learning, as a result of a continuous attempt to construct and sustain a shared and open point of view of the activity [15].

As collaboration is commonly a feature of the work environment, it must also be reflected in the design of learning activities. Johnson and Johnson [16] suggested five important key guidelines for successful collaborative learning: shortly positive interdependences, face-to-face promotion of interactions, individual accountabilities/personal responsibilities, interpersonal/small-group skills, and group processing.

In contrast cooperative learning is considered more structured, more prescriptive for teachers about classroom techniques, more directive to students about how to work together in groups, and more targeted to the public school [17]. Oxford [18] pointed out six principles for cooperative learning, which are very similar to those of Johnson and Johnson for collaborative learning [16]: positive interdependence, accountability, team formation, team size, cognitive development, social development.

Due to positive effect of collaborative and cooperative effects as stated by [19] the goal is to develop a first application that can support both. It should be usable for teachers through a structured classroom setting as well as for learners in an open and informal scenario. Because our field study took place in a classroom this chapter concentrates more on cooperative learning.

2.2   Usage of Mobile Devices in Classrooms

For an effective integration of mobile learning into a digital classroom environment, it is important for all students to have their own device equipped with wireless communication capability to conduct learning tasks [13, 20]. Various mobile devices have been used in mobile learning, such as wrist-worn devices, mobile phones, handheld computers, tablets, stylus tablet computers, and laptop computers [21]. The development of these handheld devices and wireless networking has made possible numerous new approaches to individual work and learning.

As learning becomes personal, mobile students are able to participate in collaborative learning activities when and where they want to [22]. Thus, handheld devices have been utilized to support learning, create rich learning scenarios in such technology-enriched classrooms, encourage social interaction, and support collaborative learning [23–25].

One important feature of social collaboration is intimacy, which depends mainly upon nonverbal cues such as eye contact, miming, and smiling. A mutual gaze moderates interpersonal distance and the sense of intimacy. In addition, gesticulated interaction frequently takes place along with verbal utterances in meaningful processes, resulting in meaning creation.

Scott analyzed the nonverbal interpersonal interactions of group members during discussion, to assess whether the current collaborative learning environment setting is beneficial to discussion or not. Usefulness interaction is a critical factor for success of collaborative learning [16].

Collaborative activities have gained significant attention among educators for improving student learning [25]. Instead of passively receiving knowledge from teachers alone, students can engage in collaborative activities and knowledge construction activities during peer discussion and interaction. Therefore, it seems from high interest to bring students together by connecting their devices.

2.3   Games Facilitating Cooperative and Mobile Learning

Games are often seen as a good promoter of learning activities or the other way round; the benefits of learning through games are numerous [26]. According to Malone [27, 28] three characteristics are essential for games: challenge, curiosity, and fantasy. A game must have a clear, achievable goal, and feedback must be appropriate for the users. Furthermore, the game has to be challenging and not predictable. Based on this gaming theory, several learning games have been developed [29].

The main idea of introducing games for learning is that learning happens indirectly, as a kind of side effect of playing the game. Even the learning effect might be much higher because learners are in an emotional and motivated situation [30–32]. Therefore, it can be assumed that a game design will have a positive influence to the learning behavior of the school children.

Nowadays, there are many multiplayer games with connecting mobile devices available [25], but we are not able to find a single learning game among them; at least not within the German iTunes Store (10/2012) or related research on this certain topic [33].

2.4   Cooperative Mobile Learning Within Classrooms

Building on the trend that more and more mobile devices are available in classroom and on the attractiveness and importance of cooperative and collaborative peer learning, cooperative mobile learning within classrooms comes strongly to the fore: Thus, the collaboration of students with their peers through different technological devices, and the influence of these devices on interaction among students, become important research issues.

Studies have validated experiments which help students to collaborate and exchange information through handheld devices as well as providing opportunities to interact with each other using these devices for supporting learning, thus facilitating cooperative learning, e.g., students engaged in collaborative learning through face-to-face communication on a social network with the support of handheld devices by a wireless network [24].

Nevertheless, the idea of a cooperative learning app for connected mobile devices for young learners is a new one. As pointed out before, there are no known other applications in the German iTunes Store supporting learning through connected devices.

Similar ideas are carried out in different research studies: First, children should draw together one sketch [34] or second find predefined words by collecting letters cooperatively [35].

2.5   Lessons Learned from iPhone/iPads Learning Apps Development at TU Graz

TU Graz has an established tradition of research and development of mobile applications. The first lecture on iPhone development took place in 2010 with more than 100 participants. Since then, more than 100 iPhone/iPad apps have been uploaded to the iTunes Store and offered for free by TU Graz. Therefore, a number of workshops have been given on app design and how to achieve user satisfaction [36].

At the same time the Department of Social Learning as well as the Institute for Information Systems and Computer Media (IICM) at the TU Graz started a mLearning initiative with the goal of developing educational apps and finding out how these apps change the way we learn. Each of the apps developed addresses a specific learning goal as described in the Apple Definition Statement (ADS): According to ADS, every app has to claim three major requirements (description of unique selling point, audience, and solution).

All developed educational apps are available at the website http://app.tugraz.at. It is worth noting that the target audience of these applications is a range from preschool-aged children toward students at universities.

The following is an example of one of these apps and the research on its use within higher education: Ebner and Billicsich [37] gave a first impression on how smartphones can change the traditional lecturing through digital documentation of teaching and learning events. Similar to today's well-known application "Evernote," students are able to collect photos, notes, hyperlinks, or videos of a lecture and store them online to augment the learning process.

Field research is often undertaken to evaluate an app, including its usage and effects, under the guidance of the Department of Social Learning. For example, a class of 9-year-old school children was observed to find out how they approach their iPad usage, as well as to look on the outcome of their usage within the classes.

In her study, Huber [38] pointed out how iPads can enhance the traditional classroom. According to her data, apps facilitating creativity were used more intensely [38]. Follow-up research showed how iPads help to individualize education with a special focus on language teaching [39] or how tablet computers change the way we teach—in music classrooms, for example [40].

These studies have shown that applications designed for children have certain special requirements. For example, the rotation feature must be used carefully, and the concept of file handling should be simplified as well, as children must be aware of hierarchy, position, and auxiliary views such as bars in order not to get lost. Huber and Ebner [41] published a list of such requirements for mobile learning apps aimed at children. From our point of view, the two main omissions among iPad apps used in the classroom are:

- Lack of feedback for teachers: When school children work and learn with different apps the face-to-face education is individualized. But the teacher needs to know how the child performed, how many exercises done or learning goals have been achieved. Yet mobile applications are stand-alone programs and can be done on their own, without a feedback channel for a supervisor or teacher. There is no way for the teacher to get an overview of the current performance of an individual, or of the whole class. Therefore, a central web-based infrastructure is necessary as well as a user and class management. Furthermore, an automated analysis of the outcomes is needed. Our attempts show the great potential of such learning analytics and central feedback applications [42].

- Lack of collaboration: Current mobile applications are able to enhance individualization and personal learning, as children are enabled to learn according to their particular needs and at their own pace. Developers of mobile learning apps take no specific note of cooperation and collaboration issues. This stands in contrast to the development of gaming apps, where cooperation features and connections of multiple devices are trendy and common.

In our research study we would like to particularly address the issue of lack of cooperation and collaboration by designing a cooperative/collaborative learning app running on multiple mobile devices of the learners with a special focus to math, respectively, STEM education [43].

3   Goal of the Study and Research Question

In our work we want to build on these insights and plan to develop and test a learning app for young learners, with special attention to cooperative peer learning, aspects of the games, and usage of the personal device. Additionally, we aim to bring cooperative learning with connected mobile devices into the classroom setting.

In other words, one must consider whether the connection of students' mobile devices impacts a specific learning goal. Therefore, the overall research question addressed in this work is: How can collaborative learning can be advanced in classrooms using the mobile devices of the school children?

In particular our main goal is to develop a first math app for school children that bring learners together through their connected devices. Therefore, following hypotheses (HY) have been carefully worded:

- HY1: The developed app is able to assist cooperation between learners.
- HY2: The developed app motivates learners to play the game again.
- HY3: The developed app is easy to use for learners.

4   Research Design

As there are no existing collaborative learning games, as mentioned in the first part of this paper, our research design is twofold:

- In the first place, an appropriate app has to be developed, following the principles of a cooperative classroom setting. Therefore, we describe the design and development of the app "MatheBingo". We strongly follow the approach of *prototyping* as described by [44, 45]: identifying basic requirements, development of a working prototype, implementation, usage, and revision.
- Second, an initial evaluation of the app was needed to get insights into its usage. For this, we used a *participatory observation* in a real classroom setting in an Austrian iPad class (first grade in a primary school). After a short introduction into the idea and rules of the app, the children were randomly separated into groups of four. Because of technical problems, the groups took turns playing the game. The children used the app for about 60 min on their own, assisted by the teacher and a research assistant as well as watched (and commented on) by their peers. The two participating observers were asked to make notes of their observations during that phase. Afterwards, the observers wrote a report about their observations, focusing on cooperative usage and technological aspects of the app's usage. After the math lesson, a short feedback from the children was gathered to understand their views on the application.

5 Concept and Design of the iPhone/iPad App "Mathebingo"

In the following paragraphs we describe the concept, technological considerations, the user interfaces, the rules of the game, and its expected usage. The main goal is to get an idea about the game play and how the app is working. Technical details will not be carried out.

5.1 General Concept

The general idea (ADS of the application) of "MatheBingo" was to develop a learning game in the field of math (ADS: solution) for school children in the age of 6–8 (ADS: audience) with a special focus on collaboration (ADS: differentiator). So the application should allow children to connect with each other and to work together on a learning goal [46]. The game design should help to motivate the learners to use the learning applications again and again.

The idea of the game, called MatheBingo, is rather simple. Everyone knows about how to play Bingo: Each player gets a card with randomized numbers and crosses a number out when the game leader calls it. The first player with crossed out numbers all in a row wins the game. Our application follows these rules, but in our case the cards are inputted on each individual device (in our particular case on iPads) and the leader of the game is the game table on an iPad.

Instead of just calling out a number, it requires a simple calculation appropriate to the knowledge level of the pupils. So the game table asks for simple summations or subtractions, which must be calculated by the children and afterward the result can be marked on their devices, if the appropriate number is available.

5.2 Technical Framework

Building on the specialized know-how of the sector of iPhone/iPad development, it was decided to program a so-called universal app running on both devices. ObjC is used for the programming language. The app must operate with at least iOS 5 and higher. In addition to the official iPad Human Interface Guidelines [47], the extended guidelines mentioned above considering the mLearning focus [41] must also be taken into account.

The fact that children do not have any concept of how Wi-Fi or Bluetooth works is one of the major challenges of this application. With other words necessary technologies have to work in the background without needing additional information from the learner.

5.3 Rules of the Game and Expected Usage

In accordance with the Bingo game concept, "Mathebingo" is a competitive app where the player who has the first complete row wins. Therefore, this concept does not fulfill the common idea of a "group learning goal" or "common goal", but nevertheless, this concept has an important feature of cooperative learning: Within the so-called "unlimited" option, all players have as much time as they need to find the right number on their device. This feature resulted in more children supporting one another, especially those who needed more time, due to the fact that children in this age group played the game mostly "open" by presenting their devices (or Bingo cards) to their classmates.

A very common goal for peer learning activities is that more advanced students support the others: Proficiency in math and excitement for the game should lead to students advising others for in order to be able to go on with the game. Nevertheless, aptitude in math is not the only factor needed to win, as the winner also depends on luck for the winning arrangement of numbers on his or her bingo card. Given this factor of chance, each player is potentially able to win a game.

Figure 1 shows a typical setup for a game. In the middle the game table is placed and up to four additional devices are the bingo cards. It does not matter if the devices are iPads or iPhones, respectively, it is possible to mix them (for example, iPad is used to act as game table and up to four iPhones are the bingo cards).

5.4 The Main Interfaces of the Applications and Their Features

The learning game includes five different screens: the splash screen, the preference screen, the search screen, the game table, and the bingo card. Each screen is

Fig. 1 MathBingo with one game table (*middle*) and four playing devices (*bingo cards*)

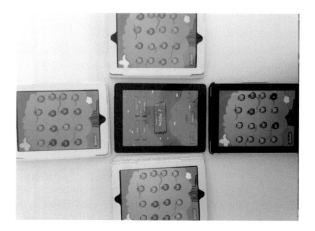

Fig. 2 Start screen

designed by a professional designer to be appropriate for school children. The splash screen (Fig. 2) is the start screen and there are three simple buttons on it.

The user taps one of the first two buttons to indicate whether the device should act as the game table or as a bingo card. Our intention is that, if available, an iPad takes the role of the game table, but this can also be done by an iPhone. Second, there is a help button, with additional instructions and guidance regarding the app's usage.

When a user chooses to act as game table, the preference screen will open. On the top the devices connected to the table are shown. Below them a snail is pictured, representing the option of changing the amount of time provided to the learner for marking the correct solution on their card. Finally, users are able to choose two additional options: the skill level (which manages the range of the required calculations) and the background music.

When the "start" button is pushed the game table is shown. The game table screen (Fig. 3) is rather simple: in the middle of the screen a calculation is provided. For example, Fig. 3 challenges the children to add 8 + 6.

If the bingo card is chosen on the splash screen (Fig. 2), the device is noted as a bingo card and the search screen is provided. This screen tells the child that his/her device is looking for a game table. If such a table is found, the device is automatically connected to it. The only requirement for a successful connection is that the devices be on the same Wi-Fi network.

The last screen is the bingo card (Fig. 4) itself. If the device has found a game table, a bingo card with 16 randomized numbers (within the defined range) placed on apples is provided. The learner can choose only one number, via a touch gesture,

Fig. 3 Game table screen

Fig. 4 Bingo card

within the given time. If he/she has solved all of the calculations in a single row (represented by red marked apples) the game is won, as signified by a big Bingo sign (Fig. 5).

Fig. 5 BINGO—end of the game

6 Using the App in the Field: Results

As described in the section "research design", a participatory observation and a plenum discussion with the children were carried out as part of our field research. In the following section, we describe the results, starting with some notes regarding the setting. Afterwards, the outcomes of the field study are clustered into three parts. First, the observations of the research assistant as well as those of the teacher are presented. Second, the feedback of children is listed, and third, technical challenges are addressed.

It must be noted that the participatory observation aims to determine whether the developed prototype works in general (technical perspective) and whether collaborative learning in general occurs (didactical perspective). The following chapters describe the general setting of the field study and the outcomes in terms of cooperation and technical issues.

6.1 The Setting of the Field Research

The research study in the field took place in an elementary school located in Vienna in late spring 2013. 22 school children between 6 and 7 years of age participated. This class is so-called "iPad-class". This means that all children get their own iPad when starting school.

Due to the fact that all of the children had been using the devices at least for more than half a year, they were comfortable using their devices. Nevertheless, the installation of the Mathe Bingo app was done beforehand to save time.

The teacher was very experienced in using iPads in the classroom. She has already implemented different applications for different purposes. Furthermore the class owns its personal blog, where the use of the tablet is documented for especially the parents and other interested people.

Finally, the technical equipment of the school was rather bad. Wi-Fi was up but tablets lost connections regularly and also the bandwidth was very low. Nevertheless, the situation is no unusual for Austrian schools especially for elementary schools.

6.2 Outcomes Related to Cooperation While Using the App

The main idea of the app was to initiate and foster cooperation and support among the children while playing the game. As sketched previously, our hope was that support would be offered by pupils who were better at math or faster at finding the right number on the device.

The following observations illustrate the cooperative aspects of the usage according to the models of Johnson and Johnson [16] and Oxford [18]:

- Face-to-face promotion of interactions: The devices lay in front of the children, so each child was able to see the bingo card of each fellow. This provided a basis for cooperation.
- Group processing: Additionally, no child played on his/her own during the observation phase. As cooperation is related to communication, it is also remarkable that there was steady communication in every group during the observation phase.
- Interpersonal/small-group skills: Another important fact was that the children helped each other a number of times—in some cases with every new assignment.
- Positive interdependences: As mentioned in the report, the observers also repeatedly describe individual children looking at the game of another group and helping them finding correct solutions. This occurred especially frequently when the child had a visiting, nonplaying role.
- Individual accountabilities/personal responsibilities: Last, but not least, the observers mentioned that even children with lower level mathematical skills were motivated to play the game and engage with topic. The observers see the fact that their chances of winning are equal to those of the high-performing children as a reason for this.
- Another valuable insight builds on a mistake: Some assignments within the app were too difficult for the pupils' level of knowledge. Surprisingly, the pupils tried to solve the challenging assignments on their own, discussed their solutions

and managed to find the correct answers. This should be taken into account as a future didactical strategy in offering some challenging calculations to the learners.

To sum up these results concerning the cooperation, the concept and goals of providing such an app for cooperation were realized. Additionally, the observers reported that the children in general seemed to enjoy finding the right answer on their cards and expressed their enthusiasm the moment the color of an apple changed to red (for the correct solution).

The children even invented a special bingo result: Sometimes a child ended up with a red apple that completed two rows at once. This phenomenon was affectionately called "double bingo".

Feedback of the children: Children mentioned positively that they "love to win the game" as well as ending up with a double bingo. They showed general enthusiasm for this kind of game and said that they would use the app in the future due to its motivating character. On the negative side, the children reported technical problems, as will be described in the next paragraph.

Building on this problem, they said that they could not play as often as they wanted. One child mentioned that the apple was not changing the color to red even though it was touched correctly. Some children were annoyed by the loud victory celebrations.

6.3   Summary of Technical and Other Challenges

As the app is also a technical innovation, technical aspects are important. In our case our field research unfortunately showed technical problems that we had not found in our prior tests: Sometimes the searching devices did not find the game table and sometimes the connection simply got lost.

The children had serious problems finding and connecting the devices within their five groups. Recognizing this, the teacher had to decide to let each group play one at a time, which then worked well. Concerning the application itself, the predefined knowledge levels did not match the usual proven method suggested from the school literature (especially regarding subtractions from 1 to 30). Finally, sometimes the touch screen did not react in time, especially when children tried to mark the right solution.

7   Discussion

In this section the methodology and findings of the observation are discussed. As is typical for this kind of development and initial field research, our methodology allows for general inferences and concrete feedback for the future development of the app and usage within classroom. Deeper insight into learning outcomes will require additional and more-experimental settings (which are currently very difficult to obtain). Such general findings concerning this learning app are. Nevertheless, we like to try to answer our HY:

- HY1: *The developed app is able to assist cooperation between learners.* The field study showed that the application enhances communication and cooperation between the children. In particular the teacher mentioned that the children solved math problems they had not yet learned about during playing the game mainly through conversation.
- HY2: *The developed app motivates learners to play the game again.* The even chances of winning the learning game seems to be a key success factor of this app, as it does not favor pupils with better math competence and motivates even children who are poor at math. This seems to be the basis for the facilitation of such a highly communicative dynamic and the children's motivation to play the learning app again: The researchers observed that this leads to a long-lasting motivation doing the game again and again. In other words, the gaming factor increases the reuse of the app dramatically as opposed to pure learning apps.
- HY3: *The developed app is easy to use for learners.* The technical solution still needs enhancements: Sometimes it takes a very long time (180 s) to get connected, and sometimes the connection simply breaks down. The observation also pointed out that children were not able to connect to each other by themselves and that assistance was necessary. Bearing in mind that they have no understanding of Wi-Fi and other similar technologies, there is some additional training necessary during the process. But all these technical issues are solvable. In general the app works and children have no further problems during playing Mathe Bingo.

Last, but not least, future work, including additional field studies and experiments, should be arranged to gain deeper insights into both the communication of the children and the learning outcomes of the game. Future research should be undertaken concerning the individual and the group learning outcomes, as well as different possible settings (e.g., rotating groups), and apps on other topics.

8 Conclusion

Communication and cooperation are key factors in successful learning. Nowadays, we have the chance to use mobile technologies within classroom and to connect them on a network. Nevertheless, there have been only few research studies done concerning this issue, mainly because there are some technical restrictions.

Within this publication a trailblazing collaborative learning app for mobile devices (iPhone/iPad) has been developed using Wi-Fi connections among several different devices. The follow-up field experiment indicates that children are highly motivated to use these kinds of applications. Quite obviously, the game's setting and its fostering of communication seem to equip it to motivate even poorer mathematicians in the class to engage with math and to learn. The application addresses all of the guidelines set out by Johnson and Johnson [16], as we have discussed, and the children clearly enjoy playing with it.

In summary, the gaming environment and communication about the problem at hand, as well as collaboration through mobile devices, is a very powerful combination. From our perspective, this first trial is very promising, and if the app is developed according to the technical capabilities of new mobile operating systems (in our case iOS7), usability and HCI guidelines with a special focus on children, as well as a theoretical pedagogical approach such as that of Johnson & Johnson tablets can enhance the quality of education in today's classrooms.

Mobile technologies have the capacity to change the landscape of our classrooms due to the fact that individualized, personalized, and collaborative learning settings can supported by appropriate applications. At the same time, it must be clearly stated that further research will be necessary to obtain more in-depth insights into how children benefit from such activities and how learning can be improved by such collaborative tools.

In conclusion, we hope to inspire others to devise and share classroom observations and experiments, or to develop similar apps. Bearing in mind that also wearable devices hit now the market it will become more and more necessary to enhance our research studies in order to guarantee that appropriate didactical scenarios are available. Learning with and through technologies is able to improve education, therefore we have to take a close look in which way.

References

1. O'Reilly, T.: What is web 2.0? Design patterns and business models for the next generation software. Commun. Strat. **65**, 17–37 (2007)
2. Downes, S.: E-learning 2.0. ACM eLearn Mag. (2005)
3. Ebner, M.: E-learning 2.0 = e-learning 1.0+ web 2.0? In: The Second International Conference on Availability, Reliability and Security, ARES 2007, pp. 1235–1239. IEEE (2007)
4. Augar, N., Raitman, R., Zhou, W.: Teaching and learning online with wikis. In: Atkinson, R., McBeath, C., Jonas-Dwyer, D., Phillips, R. (eds.) Beyond the Comfort Zone: Proceedings of the 21st ASCILITE Conference, pp. 95–104. Perth, 5–8 Dec 2004
5. Safran, C., Ebner, M., Garcia-Barrios, V.M., Kappe, F.: Higher education m-learning and e-learning scenarios for a geospatial wiki. In: E-Learn—World Conference on E-Learning in Corporate, Government, Healthcare, and Higher Education (2009)
6. Farmer, J., Bartlett-Bragg, A.: Blogs @ anywhere: high fidelity online communication. In: Proceeding of ASCILITE 2005: Balance, Fidelity, Mobility: Maintaining the Momentum? pp. 197–203 (2005)
7. Holzinger, A., Kickmeier-Rust, M., Ebner, M.: Interactive technology for enhancing distributed learning: a study on weblogs. In: Proceedings of HCI 2009. The 23nd British HCI Group Annual Conference, Cambridge, London, pp. 309–312 (2009)
8. Campell, G.: There's something in the air —podcasting in education. Educause Rev. **2005**, 33–46 (2005)

9. Nagler, W., Saranti, A., Ebner, M.: Podcasting at TU Graz: how to implement podcasting as a didactical method for teaching and learning purposes at a university of technology. In: Proceeding of the 20th World Conference on Educational Multimedia, Hypermedia and Telecommunications (ED-Media), pp. 3858–3863 (2008)
10. Maierhuber, M., Ebner, M.: Near field communication—which potentials does NFC bring for teaching and learning materials?. Int. J. Interact. Mobile Technol. 7(4), 9–14 (2013)
11. Lehner F., Nosekabel H.: The role of mobile devices in e-learning—first experience with a e-learning environment. In: Milrad, M., Hoppe, H.U., Kinshuk (eds.) IEEE International Workshop on Wireless and Mobile Technologies in Education, pp. 103–106. IEEE Computer Society Press, Los Alamitos, CA (2002)
12. Curtis, M., Luchini, K., Bobrowsky, W., Quintana, C., Soloway, E.: Handheld use in K-12: a descriptive account. Proceedings of IEEE International Workshop on Wireless and Mobile Technologies in Education (WMTE), pp. 23–30. IEEE Computer Society Press, Los Alamitos, CA (2002)
13. Liang, J.K., Liu, T.C., Wang, H.Y., Chang, L.J., Deng, Y.C., Yang, J.C., Chou, C.Y., Ko, H.W., Yang, S., Chan, T.W.: A few design perspectives on one-on-one digital classroom environment. J. Comput. Assist. Learn. 21, 181–189 (2005)
14. Dillenbourg, P.: Introduction: what do you mean by collaborative learning? In: Dillenbourg, P. (ed.) Collaborative Learning: Cognitive and Computational Approaches, pp. 1–19. Elsevier, Oxford, UK (1999)
15. Vygotsky, L.S.: Mind in Society: The Development of Higher Psychological Processes. Harvard University Press, Cambridge, UK (1978)
16. Johnson, D.W., Johnson, R.T.: Learning Together and Alone, 4th edn. Allyn and Bacon, Needham Heights, MA (1994)
17. Matthews, R.S., Cooper, J.I., Davidson, N., Hawkes, P.: Building bridges between cooperative and collaborative learning. Change 27, 35–40 (1995)
18. Oxford, R.L.: Cooperative learning, collaborative learning, and interaction: three communicative strands in the language classroom. Modern Lang. J. 81(4), 443–456 (1997)
19. Roger, T., Johnson, D.W.: Cooperative learning. Two heads learn better than one. Transforming education (IC#18), online available: http://www.context.org/iclib/ic18/johnson/, July 2015
20. Chan, T.W., Roschelle, J., Hsi, S., Kinshuk, Sharples M., Brown, T., Patton, C., Cherniavsky, J., Pea, R., Norris, C., Soloway, E., Balacheff, N., Scardamalia, M., Dillenbourg, P., Looi, C.K., Milrad, M., Hoope, U.: One-to-one technology enhanced learning: an opportunity for global research collaboration. Res. Practice Technol. Enhanced Learn. 1, 3–29 (2006)
21. Sharples, M., Beale, R.: A technical review of mobile computational devices. J. Comput. Assist. Learn. 19(3), 392–395 (2003)
22. Looi, C.K., Seow, P., Zhang, B., So, H.J., Chen, W., Wong, L.H.: Leveraging mobile technology for sustainable seamless learning: a research agenda. Br. J. Edu. Technol. 41(2), 154–169 (2010)
23. Roschelle, J., Pea, R.: A walk on the WILD side: how wireless handhelds may change computer-supported collaborative learning. Int. J. Cogn. Technol. 1(1), 145–168 (2002)
24. Zurita, G., Nussbaum, M.: Computer supported collaborative learning using wirelessly interconnected handheld computers. Comput. Educ. 42(3), 289–314 (2004)
25. Liu, C.-C., Kao, L.-C.: Do handheld devices facilitate face-to-face collaboration? Handheld devices with large shared display groupware to facilitate group interactions. Comput. Assist. Learn. 23(4), 285–299 (2007)
26. Mann, B.D., Eidelson, B.M., Fukuchi, S.G., Nissman, S.A., Robertson, S., Jardines, L.: The development of an interactive game-based tool for learning surgical management algorithms via computer. Am. J. Surg. 183(3), 305–308 (2002)
27. Malone, T.W.: What makes things fun to learn? Heuristics for designing instructional computer games. In: Proceedings of: 3rd ACM SIGSMALL Symposium and the First SIGPC Symposium on Small systems, pp. 162–169 (1980)

28. Malone, T.W.: Heuristics for designing enjoyable user interfaces: lessons from computer games. In: Proceedings of Conference on Human Factors in Computing Systems, Gaithersburg (MD), pp. 63–68 (1982)
29. Ebner, M., Holzinger, A.: Successful implementation of user-centered game based learning in higher education: an example from civil engineering. Comput. Educ. **49**(3), 873–890 (2007)
30. Brehm, J.W., Self, E.A.: The intensity of motivation. Annu. Rev. Psychol. **40**, 109–131 (1989)
31. Holzinger, A., Maurer, H.: Incidental learning, motivation and the tamagotchi effect: VR-Friends, chances for new ways of learning with computers. In: Proceedings of Computer Assisted Learning, CAL 99, p. 70. London (1999)
32. Holzinger, A., Pichler, A., Almer, W., Maurer, H.: TRIANGLE: a multi-media test-bed for examining incidental learning, motivation and the Tamagotchi-effect within a game-show like computer based learning module. In: Proceedings of Educational Multimedia, Hypermedia and Telecommunication 2001, pp. 766–771. Tampere, Finland (2001)
33. Schön, S., Wieden-Bischof, D., Schneider, C., Schumann, M.: Mobile Gemeinschaften. Erfolgreiche Beispiele aus den Bereichen Spielen, Lernen und Gesundheit. Salzburg: Salzburg Research (2011)
34. Spitzer, M., Ebner, M.: Collaborative learning through drawing on iPads. In: Proceedings of World Conference on Educational Multimedia, Hypermedia and Telecommunications 2015, pp. 633–642. AACE, Chesapeake, VA (2015)
35. Ebner, M., Kienleitner, B.: A contribution to collaborative learning using iPads for school children, European Immersive Education Summit, Vienna, pp. 87–97 (2014)
36. Ebner, M., Kolbitsch, J., Stickel, C.: iPhone/iPad human interface design. In: Human-Computer Interaction in Work and Learning, Life and Leisure, pp. 489–492 (2010)
37. Ebner, M., Billicsich, T.: Is the iPhone an ubiquitous learning device? A first step towards digital lecture notes. In: Ubiquitous Learning—Strategies for Pedagogy, Course Design and Technology, pp. 137–151 (2011)
38. Huber, S.: iPads in schools—blessing or curse? Unpublished term paper at Graz University of Technology (2011)
39. Huber, S.: iPads in the classroom. Masterthesis at Graz University of Technology, Book on Demand GmbH., Norderstedt, German (2012). Retrieved Feb 2013, from: http://itug.eu
40. Frühwirth, A.: Innovativer Technologieeinsatz im Musikunterricht, Master thesis at Graz University of Technology (2013)
41. Huber, S., Ebner, M.: iPad human interface guidelines for m-learning. In: Berge, Z.L., Muilenburg, L.Y. (eds.) Handbook of Mobile Learning, pp. 318–328. Routledge, New York (2013)
42. Ebner, M., Schön, M.: Why learning analytics in primary education matters!. Bull. Tech. Committee Learn. Technol. Karagiannidis, C., Graf, S. (eds.) **15**(2), 14–17 (2013)
43. Ebner, M.: Mobile learning and mathematics. Foundations, design, and case studies. Crompton, H., Traxler, J. (eds.) Routledge, New York, pp. 20–32 (2015)
44. Alavi, M.: An assessment of the prototyping approach to information systems development. Commun. ACM **27**(6), 556–563 (1984)
45. Larson, O.: Information systems prototyping. In: Proceedings Interest HP 3000 Conference, Madrid, pp. 351–364 (1986). www.openmpe.com/cslproceed/HPIX86/P351.pdf. Accessed July 2015
46. Nakagawa, H., Sugiyama, R., Sato, Y.: The development and evaluation of a collaborative learning tool for the visualization of thought and promoting exchange amongst children. In: World Conference on Educational Multimedia, Hypermedia and Telecommunications, Poster. AACE, Chesapeake, VA (2013)
47. iPad Human Interface Guidelines.: Retrieved Feb 2013 from Apple Inc. (2011). http://www.scribd.com/doc/61285332/iPad-Human-Interface-Guideline

Prospective Teachers—Are They Already Mobile?

Süleyman Nihat Şad, Özlem Göktaş and Martin Ebner

Abstract This research study investigated the prospective teachers' purposes of using mobile phones and laptops, as well as the significant differences across genders and grades. Furthermore, the frequency of connecting to Internet via both mobile devices was investigated comparatively. The study was designed based on cross-sectional survey and casual-comparative methodologies in order to first determine specific characteristics of the relevant population, and to determine the possible causes for differences in terms of variables investigated. A total of 650 prospective Turkish teachers participated in the study. The results point out that, compared to mobile phones, laptops were used more frequently for various purposes, particularly the educational ones. However, in-class use of both laptops and mobile phones for educational purposes was not very common. Mobile phones were used less for educational purposes, but more for communication and entertainment purposes. Though there were statistically significant differences in terms of some purposes, given the lack of practical significance, both male and female prospective teachers can be said to use mobile phones and laptops for various purposes with similar frequencies. The same was also true for the grade variable: all prospective teachers from first to fourth years used mobile phones and laptops for various purposes with similar frequencies in practice. The present study also revealed that, for prospective teachers, connecting to the Internet via mobile phones is not very common and even significantly less common than doing so via laptops.

S.N. Şad (✉)
İnönü University, Curriculum and Instruction, 44280 Malatya, Turkey
e-mail: nihat.sad@inonu.edu.tr

Ö. Göktaş
Ministry of National Education, Sumer Secondary School, 44100 Yesilyurt, Malatya, Turkey
e-mail: ozlemgoktas44@hotmail.com

M. Ebner
Educational Technology, Graz University of Technology, Graz, Austria
e-mail: martin.ebner@tugraz.at

© Springer International Publishing Switzerland 2016
A. Peña-Ayala (ed.), *Mobile, Ubiquitous, and Pervasive Learning*,
Advances in Intelligent Systems and Computing 406,
DOI 10.1007/978-3-319-26518-6_6

139

The findings in general suggested a need to raise awareness among prospective teachers about the mobile learning potential of mobile phones in general and in-class use of laptops in particular.

Keywords Mobile learning · M-learning · Pre-service teachers · Laptops · Mobile phones

Abbreviation List

CK	Content knowledge
e-learning	Electronic learning
m-learning	Mobile learning
m-phone	Mobile phone
PC	Personal computer
PCK	Pedagogical content knowledge
PDA	Personal digital assistant
PK	Pedagogical knowledge
SMS	Short message service
TCK	Technological content knowledge
TK	Technological knowledge
TPCK	Technological pedagogical content knowledge
TPK	Technological pedagogical knowledge

1   Introduction

Since the 1980s technologies have become an important agenda in educational discourse mainly because digital technologies became widely available and learning how to use them in teaching became a must, changing the nature of classroom and instruction [1]. This caused to redefine the competences required by teaching profession so as to include the technological competences. Today, both in-service and pre-service teachers are expected to know how to integrate technology into teaching and learning processes [2–6].

One recent approach regarding the integration of technology into teacher education and teacher professional development context is the framework of "Technological Pedagogical Content Knowledge (TPCK)". Proposed by Mishra and Koehler [1], the conceptual framework of TPCK was intended to provide a theoretical grounding for educational technology by extending "Shulman's formulation of 'pedagogical content knowledge' to include the phenomenon of teachers integrating technology" (p. 1017). Though a rather new concept, TPCK has become a commonly referred framework in defining a teacher's technology integration competences especially in terms of teacher training context [5, 7–11].

TPCK framework is based on three main components (content knowledge—CK, pedagogical knowledge—PK and technological knowledge—TK) and three pairs of knowledge intersection (pedagogical content knowledge—PCK, technological content knowledge—TCK and technological pedagogical knowledge—TPK) and one triad (technological pedagogical content knowledge—TPCK) [1]. To Koehler and Mishra [12] in the model technology has an integral part requiring the teachers to "accomplish a variety of different tasks using information technology and to develop different ways of accomplishing a given task" (TK) (p. 15), "have a deep understanding of the manner in which the subject matter (or the kinds of representations that can be constructed) can be changed by the application of technology" (TCK) (p. 16), seek "forward-looking, creative, and open-minded" ways of using "technology, not for its own sake, but for the sake of advancing student learning and understanding" (TPK) (p. 17) and "understand the representation of concepts using technologies; pedagogical techniques that use technologies in constructive ways to teach content; knowledge of what makes concepts difficult or easy to learn and how technology can help redress some of the problems that students face; knowledge of students' prior knowledge and theories of epistemology; and knowledge of how technologies can be used to build on existing knowledge and to develop new epistemologies or strengthen old ones" (TPCK) (pp. 17–18).

Especially in the Turkish context, researches show that either teachers or prospective teachers have not reached the desirable level in terms of technology integration [13–16]. In a comprehensive study, Tezci [17] surveyed 1540 teachers from 330 primary schools in 18 cities in four geographical regions of Turkey and found out teachers lack both favourable attitudinal input and technical capabilities to integrate information technologies into education. However, it is evident that today instructional environments arranged by teachers who have poor computer literacy, cannot use instructional technologies like interactive whiteboards, tablets, etc. adequately, are not aware of the importance of integrating technology into their instruction, and are reluctant to learn and use new or changing technologies that are not productive or effective places to learn [6]. Thus, raising teachers with skills regarding technology integration is a must today. Russell et al. [18], for example, suggest that pre-service and in-service teacher education programmes may encourage teachers how to use "technology to deliver instruction, to prepare for instruction, to accommodate instruction, to communicate with others in and out of the school, and to direct students to use technology for specific instructional purposes" (p. 307).

It is difficult to completely list and define the spectrum of technologies to deliver instruction or to direct students to use for specific instructional purposes. However, one recent domain of instructional technologies that has come into prominence recently seems to be mobile learning, which has evolved from e-learning and computer-supported collaborative learning [19]. As presented below in the literature review, mobile technologies like laptops, smartphones, tablets, iPods, etc., have become very prevalent especially among young population and learning with these mobile devices has its promises and limitations in terms of education [20, 21].

Despite their ubiquity and flexibility, using these devices for educational purposes had little room in educational sectors and developments seem to be more about the design of the tools than ensuing learning by them [22]. Moreover, although mobile phones are gradually transforming into handheld computers, university students may still perceive mobile phones beneficial for fun and communication purposes rather than education [23] or prospective teachers may not believe in the potentials of m-phones as m-learning tools [24].

From the TPCK perspective it can be claimed that future teachers should also learn how to integrate the mobile devices into teaching learning process. This is because when they become teachers they will be either teaching/having feedback about their subject matter using mobile devices like laptops, tablet PCs, smartphones, etc., or teaching their learners the strategies about using mobile devices for learning. Thus, we believe it is necessary to understand the actual situation about prospective teachers' use of mobile devices for different purposes including the educational ones.

Describing how prospective teachers use these mobile tools is considered to be critical in understanding to what extent mobile learning has been adopted by prospective teachers. Thus, this chapter starts with a condensed review of literature about mobile technologies in general and mobile learning in particular with its promises and limitations, and educational and non-educational uses of mobile tools especially among higher education students. Next, the chapter presents the results of a cross-sectional and casual-comparative study, in which frequencies of Turkish prospective teachers to use laptops and mobile phones for different purposes (for *education, entertainment, communication/information* and *others*) were described, compared. Also results about differences according to gender and grade were presented. Lastly, frequencies of prospective teachers' connecting to Internet via mobile phones and via laptops were described and compared in the study. The chapter ends with the discussion of the research findings, followed by conclusions and implications in terms of teacher training.

2   Literature Review

To present a clear outline of the relevant literature, we have summarized the literature review below under subtitles. The introductory subtitle *Prevalence of mobile technology* gives brief info about the prevalence of mobile devices especially among young population. The subtitle *Mobile learning* presents the background, definition and current situation about mobile learning. The subtitle *Promises and limitations* summarizes the literature about the advantages and the limitations of mobile learning. Last subtitle *Mobile learning in higher education* presents a review of literature about educational and non-educational uses of mobile tools especially among higher education students.

2.1 Prevalence of Mobile Technology

Fast-growing technologies change our lives in many aspects including work habits, how we access to information, how we socialize or learn, etc., Today, a large spectrum of mobile devices including cellular phones, smartphones, mp3 and mp4 players, iPods, digital cameras, data travellers, personal digital assistants (PDAs), netbooks, laptops, tablets, e-readers, etc., have been introduced into our lives very smoothly [25–27].

With these mobile devices, the world has become not only a global village, but also a *"mobigital virtual space* where people can learn and teach digitally anywhere and anytime"* (p. 606) [24]. These mobile devices have been adopted very rapidly and have become popular [27, 28] especially among younger people such as the university students [27, 29–31]. To illustrate, even in developing countries, over 90 % of young adults from 16 to 24-year-olds have a mobile phone [27]. Cheon et al. [29] reported in their study that 152 college students out of 177 (86 %) had mobile devices. All 107 graduate and undergraduate students participating in Corbeil and Valdes-Corbeil's [32] study had a mobile or smartphone while 92 % had laptops, which was even more than that of faculty members (83 %). In Finland, even a decade ago, about 98 % of university students had a mobile phone [33].

2.2 Mobile Learning

With the advancement of mobile technologies in daily life, learning via mobile devices has gradually become a widely investigated research subject in education [34]. Hence, the new term 'mobile learning' or 'm-learning' was introduced as the next phase of electronic learning (e-learning) in the early 2000s [35]. Pereira and Rodrigues [19], for example, argue that mobile learning has evolved from e-learning and computer-supported collaborative learning. Though the first definitions called it "mobile e-learning" (p. 1345) [27], it began to be known by its current name, mobile learning or m-learning, which is in and of itself "a highly popular multidisciplinary study field around the world" (p. 202) [36].

M-learning is a relatively new phenomenon and a sound theoretical basis has not been developed yet [22]. Thus, despite its popularity an exact and well-accepted definition of m-learning has not been done yet. It can be broadly defined as "learning that happens anywhere, anytime." (p. 261) [26], or can be defined with a special emphasis on the need for using certain devices, i.e. "the use of portable devices with Internet connection capability in education contexts" (p. 639) [37] or "learning that happens on any pervasive computing devices" (p. 6) [34].

As it can be understood from these different definitions, mobile learning is regarded both as an extensive self-directed learning activity and as an educator-led planned activity. Thus, it is wise to distinguish these two practices as did by Wang and Shen [38], who define mobile learning that "occurs under management of a teacher (and generally in a purposefully built environment)" as formal and the kind that "occurs under self-management of the learner and in ad hoc environments" as informal (p. 563).

In a first final analysis, mobile learning can be defined as an extensive self-directed or educator-led planned process of learning using any mobile information and communication devices. According to a review by Wu et al. [39], mobile phones and PDAs are the devices used most commonly for mobile learning. Other mobile devices commonly used for mobile learning include laptops, smartphones, netbooks, electronic dictionaries, portable multimedia players, and iPods [26, 36, 30]. Although some [30] confine m-learning to "handheld or palmtop devices" (p. 592), laptops are also considered as a common mobile learning device [26, 40–42].

2.3 Promises and Limitations

Formal or informal, mobile learning has been reported to bring about advantages in education as exemplified below. First of all, its potential to provide instant access to information resources is commonly highlighted in several studies [34, 43, 44]. Accordingly, the removal of time and space limitations is believed to characterize mobile learning [26, 34, 35, 45–48], and is highlighted to be the most important feature by its users in higher education [29, 49].

Another commonly discussed learner-centred benefit of mobile learning is its ability to let learners learn at their own speed [26, 29]. Relevant literature highlights the interactive power of mobile learning between peers, teacher, and learners, which in turn enhances learning efficiency [29, 30, 32, 34, 49–51].

Mobile learning environments are also believed to provide new, exciting and performance-based learning opportunities [52–54] and elementary students think using mobile tools (PDAs) is more interesting than teacher-guided field trips [42]. The findings about the motivational power of mobile learning, however, are controversial. Some studies suggested that mobile learning tools increase student motivation [49, 55, 56], while some others found that mobile phones especially do not increase university students' level of motivation to learn [57]. Kinash et al. [37], on the other hand, reported that majority of students do not perceive mobile tools as having a motivational effect on learners.

Though most mobile learning studies (86 % of 164) present positive learning outcomes [39], there are also many limitations attributed to mobile learning [58]. In terms of technical limitations, limited screen size [23, 43, 57] and limited memory [59] are common problems attributed particularly to mobile phones, and limited battery life is cited as a disadvantage of laptop use [60, 61]. Data security is another limitation frequently cited in the literature [45, 62]. One of the most remarked upon limitations in the literature about mobile learning is the during-class distraction factor of mobile tools [29] especially, that of laptops [63] and phones [23]. Fried [63] for example, reported that spending considerable time with laptops for things other than taking notes during lectures interferes with students' ability to pay attention to and understand the lecture material, which in turn results in lower test scores.

As a result of the above-mentioned limitations, teachers may actually ban handphones, PDAs or portable laptops in educational settings [51]. However, it would be more advantageous to ensure that mobile tools are aligned with the course objectives so that "they make pedagogical sense" (p. 240) [51]. For instance, using mobile tools in class and school must be negotiated pedagogically. Jeng et al. [34] refer to the role of the teacher as a "mobile coacher" which involves scaffolding the learning in line with the learner's needs and abilities.

2.4 Mobile Learning in Higher Education

Higher education students in Europe prefer environments rich in multimedia images where they can actively get involved in tasks like working in groups through media sharing sites (such as Flickr or YouTube) and with a profile on social networking sites (such as MySpace or Facebook) [64]. Mobile learning is most prevalent at higher education institutions, which is followed by elementary schools [39]. There are several universities which have formally integrated mobile learning applications into their courses [27].

The educational purposes of mobile devices include either administrative ones like communicating with students for registration and administration transactions or instructional ones like delivering lesson content [65]; monitoring student progress and providing feedback to students [66]; using simulated M3G technologies in teaching physics [67]; announcing homework assignments or emailing products of group work between members [46]; providing mini lectures, student presentations, learning goal-oriented walking tours, or field visits [51]; delivering textbook-based mobile content such as reading, listening, matching and multiple choice activities [68] testing vocabulary with interactive short message service (SMS) quizzes [48]; using video recording features to evaluate learners' speaking skills [69] or finally use it for microblogging in various ways [70]. In a teacher training context, Seppälä and Alamäki [33] piloted a successful application of mobile learning where teachers and students discussed teaching issues with mobile devices, including SMS-messages and digital pictures, as a part of the supervising process.

However, as reported in several studies [26, 37, 71] mobile devices are used more for non-educative purposes than they are used for learning. These non-educative uses of mobile devices mainly include leisure time activities such as texting with friends, listening to music, surfing the web for pleasure and visiting social network services [26, 37, 72].

3   Purpose of the Study

Raising teachers with skills regarding technology integration is a must today. Teachers can create productive and effective learning environments for their students if they are willing to learn and use new or changing technologies [6]. Though it is difficult to produce a complete list of emerging technologies, one recent domain of instructional technologies seems to be mobile learning. Mobile technologies like laptops, smartphones, tablets, iPods, etc., have become very prevalent especially among young population and learning with these mobile devices has its promises and limitations in terms of education [20, 21].

Despite their ubiquity and flexibility, using these devices for educational purposes had little room in educational sectors and developments seem to be more about the design of the tools than ensuing learning by them [22]. Moreover, although mobile phones are gradually transforming into handheld computers, university students may still perceive mobile phones beneficial for fun and communication purposes rather than education [23] or prospective teachers may not believe in the potentials of m-phones as m-learning tools [24].

It can be claimed that future teachers should learn how to integrate the mobile devices into teaching learning process. This is because when they become teachers they will be either teaching/having feedback about their subject matter using mobile devices like laptops, tablet PCs, smartphones, etc., or teaching their learners the strategies about using mobile devices for learning. Thus, we believe it is necessary to understand the actual situation about prospective teachers' use of mobile devices for different purposes including the educational ones.

Describing how prospective teachers use these mobile tools is considered to be critical in understanding to what extent mobile learning has been adopted by prospective teachers. Thus the present study aimed to comparatively investigate the prospective teachers' frequency of mobile phone and laptop use for different purposes, including education, entertainment, communication/information and others. The reason to confine the research to mobile phones and laptops specifically is their ubiquitousness [26, 32, 60, 73, 74]. Also, prospective teachers' frequency of using these mobile tools for different purposes was analysed across gender and grade variables. Finally, we investigated and compared students' frequency of connecting to the Internet with these mobile tools. For this purpose, answers to following research questions were sought:

1. How often do prospective teachers use mobile phones and laptops for different purposes (e.g. education, entertainment, communication/information and others)?
2. Is there a statistically significant difference between prospective teachers' use of mobile phones and laptops for different purposes?
3. Is there a statistically significant difference between prospective teachers' use of mobile phones and laptops for different purposes by gender?
4. Is there a statistically significant difference between prospective teachers' use of mobile phones and laptops for different purposes by grade?

5. How often do prospective teachers connect to Internet via mobile phones and laptops?
6. Is there a statistically significant difference between the frequency of prospective teachers' connecting to Internet via mobile phones and via laptops?

4   Research Method

In line with these research questions above, the study was designed based on cross-sectional survey and casual-comparative methodologies in order to first determine specific characteristics of the relevant population, and to determine the possible causes for differences in terms of variables investigated [75]. Thus, we first determined frequencies of prospective teachers to use laptops and mobile phones for different purposes. Next, we compared the participants' frequencies of laptop and mobile phone use for different purposes, including the differences regarding gender and grade. Finally, the participants' frequencies of connecting to Internet via mobile phones and via laptops were described and compared.

4.1   Sampling

A total of 650 prospective teachers participated in the study from two universities in two cities in Turkey [$n = 427$ (65.7 %) in Sivas and $n = 223$ (34.3 %) in Malatya]. Sampling was done according to purposive sampling strategy, where as a part of a larger study only the students possessing both laptops and mobile phones were selected. Descriptive information about the participating prospective teachers can be seen in Table 1. Accordingly, the sample consisted of 67 (10.3 %) prospective science teachers, 58 (8.9 %) prospective social studies teachers, 68 (10.5 %) prospective primary mathematics teachers, 51 (7.8 %) prospective classroom teachers, 39 (6 %) prospective counselling and guidance teachers, 65 (10 %) prospective music teachers, 58 (8.9 %) prospective Turkish language teachers, 51 (7.8 %) prospective preschool teachers, 47 (7.2 %) prospective secondary mathematics teachers, 33 (5.1 %) prospective religion and moral teachers, 48 (7.1 %) prospective art teachers, 16 (2.5 %) prospective physical education and sport teachers, 21 (3.2 %) prospective English language teachers, 17 (2.6 %) prospective computer and technology teachers, and 11 (1.7 %) prospective special education teachers.

The national education system in Turkey is currently structured to involve a 4-year primary stage, a 4-year secondary stage, and finally a 4-year high school stage (popularly known as 4 + 4 + 4 education system). The graduates of preschool teaching department teach preschool students. Graduates of classroom teaching department teach at the first 4-year primary stage, which accepts students from

Table 1 Descriptive information about the participating prospective teachers

Variables		f	$\%$
City	Sivas	427	65.7
	Malatya	223	34.3
	Total	650	100.0
Programme	Science teaching	67	10.3
	Social studies teaching	58	8.9
	Primary mathematics teaching	68	10.5
	Classroom teaching	51	7.8
	Counselling and guidance	39	6.0
	Music teaching	65	10.0
	Turkish language teaching	58	8.9
	Preschool teaching	51	7.8
	Secondary mathematics teaching	47	7.2
	Religion and moral teachers	33	5.1
	Art teaching	48	7.4
	Physical education and sports teaching	16	2.5
	English language teaching	21	3.2
	Computer and technology teaching	17	2.6
	Special education teaching	11	1.7
	Total	650	100.0
Year	First	154	23.7
	Second	167	25.7
	Third	177	27.2
	Fourth	140	21.5
	Missing data	12	1.8
	Total	650	100.0
Gender	Female	430	66.2
	Male	220	33.8
	Total	650	100.0

66 months of age upward. Graduates of other departments mostly work at the second 4-year stage (secondary school) and partially at the last 4-year stage (high schools).

Furthermore, 154 (23.7 %) of the participants were first-year students, 167 (25.7 %) were second-year students, 177 (27.2 %) were third-year students, 140 (21.5 %) were fourth-year students, and 12 (1.8 %) did not report about their year in university. Female prospective teachers ($n = 430$) represented 66.2 % of the sample, while males ($n = 220$) represented 33.8 % of the sample.

4.2 Data Collection and Analysis

The data was collected as part of a larger study during 2011–2012 academic year. The first part of that larger study investigated the perceptions of 1087 prospective teachers about using mobile phones and laptops in education as mobile learning tools [24]. As a follow-up study, only those owning both a laptop and mobile phone were asked to respond to a second instrument about the frequency of using their mobile devices for different purposes. This instrument was a five-point (ranging between *Always* to *Never*) Likert type of scale developed to measure the prospective teachers' frequency of using laptops and mobile phones for different purposes based on the relevant literature [23, 26, 29, 30, 37, 58, 71].

The content of the instrument was validated through a qualitative pilot study and expert panel analysis. First, a pilot group of fifty prospective teachers from different departments were given a draft questionnaire including several purposes of laptop and mobile phone use. They were asked to comment on the purposes for using laptops and mobile phones included in the draft instrument and add more if any. As a result of this rather structured interviews, scale items were produced under four categories of purposes: *education, entertainment, communication/information* and *others*. Then, the draft form was consulted to the views of an expert panel involving faculty members specialized in computer and technology education and test development, and the form was finalized based on their views.

In total, 26 items under four categories of purposes were listed with two sets of responses of frequency of use (*Always* to *Never*) in two columns, one for laptops and second for mobile phones. In other words, each participant answered the same item twice: first for *laptops* and second for *mobile phone*. See Fig. 1 for the structure of the instrument and some sample items.

Prospective teachers' frequencies of using laptops and mobile phones for different purposes were analysed using mean scores ($\bar{x}$) for each item ranging from 1 referring to *never* to 5 referring to *always*. Next, participants' frequencies of using laptops and mobile phones for different purposes were compared using the non-parametric Wilcoxon test. The differences across groups, e.g. gender and year at university were tested using non-parametric Mann Whitney U and Kruskal Wallis tests, respectively. Non-parametric statistics were preferred since the comparisons were done item by item. The effect sizes were calculated using $r = \frac{z}{\sqrt{N}}$, where z is the z-score and N is the number of total observations [76]. Frequencies of

How often do you use your laptop and mobile phone for the purposes listed below? Circle 1 for *never*, 2 for *seldom*, 3 for *sometimes*, 4 for *usually*, and 5 for *always*.	Laptop					Mobile phone				
1. Recording lessons in audio format	①	②	③	④	⑤	①	②	③	④	⑤
2. Recording lessons in in video format	①	②	③	④	⑤	①	②	③	④	⑤

Fig. 1 Structure of the instrument and sample items

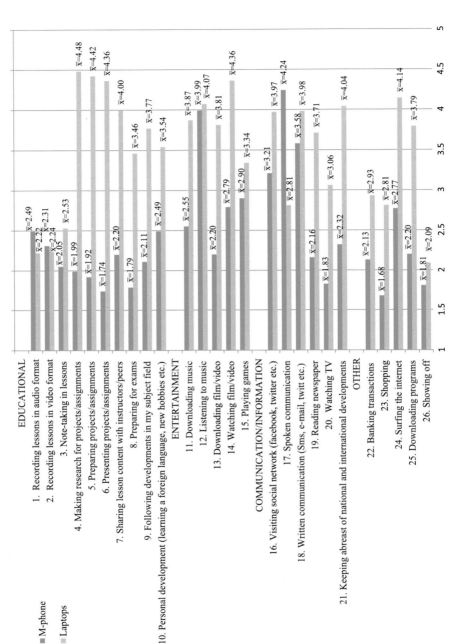

Fig. 2 Descriptive for students' frequency of using m-phone and laptop for different purposes ($n = 650$)

connecting to Internet with laptops and mobile phones were analysed using descriptive statistics (frequencies, percentages, mean scores, etc.,) and compared using non-parametric Wilcoxon test again. The significance level in inferential analysis was considered 0.05.

5 Results

Both descriptive and inferential statistical analysis yielded results regarding the research questions. The findings were presented and interpreted below under subtitles in accordance with the research questions.

5.1 Results About Purposes of Laptop and Mobile Phone Use

As it is seen in Fig. 2, prospective teachers were found to use laptops [L] mainly for educational purposes, i.e. to make research for ($\bar{x}_{L4} = 4.48$), to prepare ($\bar{x}_{L5} = 4.42$) and to present ($\bar{x}_{L6} = 4.36$) projects/assignments. Other purposes that the laptops were most commonly (*usually* to *always*) used for included *watching films/videos* ($\bar{x}_{L14} = 4.36$), *surfing the Internet* ($\bar{x}_{L24} = 4.14$), listeningto music ($\bar{x}_{L12} = 4.07$), keeping abreast of national and international developments ($\bar{x}_{L21} = 4.04$) and sharing lesson content with instructors/peers ($\bar{x}_{L7} = 4.00$).

Participants stated that they also used laptops frequently for such purposes as to perform written communication (SMS, e-mail, Twitter, etc.) ($\bar{x}_{L18} = 3.98$), to visit social networks (Facebook, Twitter, etc.) ($\bar{x}_{L16} = 3.97$), to download music ($\bar{x}_{L11} = 3.87$), to download film/video ($\bar{x}_{L13} = 3.81$), to download programmes ($\bar{x}_{L25} = 3.79$), to follow developments in their subject fields ($\bar{x}_{L9} = 3.77$), to read newspapers ($\bar{x}_{L19} = 3.71$), for personal development (e.g. learning a foreign language, new hobbies, etc.) ($\bar{x}_{L10} = 3.54$), to prepare for exams ($\bar{x}_{L8} = 3.46$), to play games ($\bar{x}_{L15} = 3.34$) and to watch TV ($\bar{x}_{L20} = 3.06$), respectively. Finally, participants were found to use laptops the least commonly for banking transactions ($\bar{x}_{L22} = 2.93$), for spoken communication ($\bar{x}_{L17} = 2.81$), for shopping ($\bar{x}_{L23} = 2.81$), to take notes during lessons ($\bar{x}_{L3} = 2.53$), to record lessons in video format ($\bar{x}_{L2} = 2.24$), to record lessons in audio format ($\bar{x}_{L1} = 2.22$), and to show-off ($\bar{x}_{L26} = 2.09$).

When it comes to mobile phones [MP], they were found to be used mainly for spoken communication ($\bar{x}_{MP17} = 4.24$). Other relatively common purposes of mobile phone use were listening to music ($\bar{x}_{MP12} = 3.99$), written communication (SMS, e-mail, Twitter, etc.) ($\bar{x}_{MP18} = 3.58$), and visiting social networks (Facebook, Twitter, etc.) ($\bar{x}_{MP16} = 3.21$). Prospective teachers were observed to use mobile phones less frequently for such purposes as playing games ($\bar{x}_{MP15} = 2.90$), watching film/video ($\bar{x}_{MP14} = 2.79$), surfing the Internet ($\bar{x}_{MP24} = 2.77$), downloading music ($\bar{x}_{MP11} = 2.55$), personal development (learning a foreign language, new hobbies, etc.) ($\bar{x}_{MP10}$ = 2.49), recording lessons in audio format ($\bar{x}_{MP1}$ = 2.49), keeping abreast of national and international developments ($\bar{x}_{MP21}$ = 2.32), recording lessons in video format ($\bar{x}_{MP2}$ = 2.31), sharing lesson content with instructors/peers ($\bar{x}_{MP7}$ = 2.20), downloading film/video ($\bar{x}_{MP13}$ = 2.20), downloading programmes ($\bar{x}_{MP25}$ = 2.20), reading newspaper ($\bar{x}_{MP19}$ = 2.16), banking transactions ($\bar{x}_{MP22}$ = 2.13), following developments in my subject field ($\bar{x}_{MP9}$ = 2.11), and note-taking in lessons ($\bar{x}_{MP3}$ = 2.05).

Finally, respectively, the least common use of mobile phones among prospective teachers was found to be for making research for projects/assignments ($\bar{x}_{MP4}$ = 1.99), for preparing projects/assignments ($\bar{x}_{MP5}$ = 1.92), for watching TV ($\bar{x}_{MP20}$ = 1.83), for showing-off ($\bar{x}_{MP26}$ = 1.81), for preparing for exams ($\bar{x}_{MP8}$ = 1.79), for presenting projects/assignments ($\bar{x}_{MP6}$ = 1.74), and for shopping ($\bar{x}_{MP23}$ = 1.68).

5.2 Comparing Frequency of Laptop and Mobile Phones Use for Different Purposes

The frequencies of using laptops and mobile phones for different purposes were compared using non-parametric Wilcoxon test for each purpose, the results of which are presented in Table 2. The results indicated that laptops were used more frequently than mobile phones in general for all categories of purposes: *education, entertainment, communication/information and others.* Accordingly, students were found to use laptops more frequently than mobile phones to take notes during lessons, to make research for projects or assignments, to prepare projects or assignments, to present projects or assignments, to share lesson content with instructors/peers, to prepare for exams, to follow developments in their subject fields, to contribute to their personal development (such as learning a foreign language, new hobbies, etc.), to download music, to download film/video, to watch film or video, to play games, to visit social network (Facebook, Twitter, etc.), to make written communication (SMS, e-mail, Twitter, etc.), to read newspapers, to watch TV, to keep abreast of national and international developments, to make banking transactions, to do shopping, to surf on the Internet, to download programmes and to show-off.

However, for two purposes the situation was reversed: *recording lessons in audio format* (item 1) and *spoken communication* (item 17), i.e. students used mobile phones more frequently than laptops for recording lessons in audio format and also students use mobile phones more frequently than laptops for spoken communication.

Moreover, for two items (i.e. "*2. Recording lessons in video format*" and "*12. Listening to music*"), there were no statistically significant differences between frequency of using mobile phones and laptops ($p > 0.05$). On the other hand most of the differences were proven to be practically significant, as well, with large effect sizes for eight of the items ($r_{(6)}$ = 0.57, $r_{(4)}$ = 0.57, $r_{(5)}$ = 0.56, $r_{(7)}$ = 0.51, $r_{(8)}$ = 0.51, $r_{(9)}$ = 0.50, $r_{(21)}$ = 0.51, $r_{(25)}$ = 0.50) and medium effect sizes for most items ($r_{(14)}$ = 0.49; $r_{(17)}$ = 0.49; $r_{(19)}$ = 0.49; $r_{(13)}$ = 0.48; $r_{(24)}$ = 0.45; $r_{(11)}$ = 0.44;

Table 2 Comparison of the frequency of laptop and mobile phones use for different purposes

Purposes	Laptop–mobile phone	N	Mean rank	Sum of ranks	Z	p	r
1. Recording lessons in audio format	Negative ranks	214	166.87	35,711.00	−5.182[a]	0.000*	0.14
	Positive ranks	114	160.04	18,245.00			
	Ties	322					
2. Recording lessons in video format	Negative ranks	170	150.97	25,665.50	−1.11[a]	0.267	0.03
	Positive ranks	139	159.92	22,229.50			
	Ties	341					
3. Note-taking in lessons	Negative ranks	90	137.49	12,374.00	−8.037[b]	0.000*	0.220
	Positive ranks	228	168.19	38,347.00			
	Ties	332					
4. Making research for projects/assignments	Negative ranks	25	86.64	2,166.00	−20.46[b]	0.000*	0.57
	Positive ranks	549	296.65	162,859.00			
	Ties	76					
5. Preparing projects/assignments	Negative ranks	19	74.03	1406.50	−20.37[b]	0.000*	0.56
	Positive ranks	540	287.25	155,113.50			
	Ties	91					
6. Presenting projects/assignments	Negative ranks	31	90.52	2806.00	−20.61[b]	0.000*	0.57
	Positive ranks	556	305.35	169,772.00			
	Ties	63					
7. Sharing lesson content with instructors/peers	Negative ranks	36	132.64	4775.00	−18.44[b]	0.000*	0.51
	Positive ranks	481	268.46	129,128.00			
	Ties	133					
8. Preparing for exams	Negative ranks	36	178.39	6422.00	−18.26[b]	0.000*	0.51
	Positive ranks	492	270.80	133,234.00			
	Ties	122					
9. Following developments in my subject field	Negative ranks	33	176.47	5823.50	−18.17[b]	0.000*	0.50
	Positive ranks	485	265.15	128,597.50			
	Ties	132					
10. Personal development (learning a foreign language, new hobbies, etc.)	Negative ranks	43	130.02	5591.00	−15.45[b]	0.000*	0.43
	Positive ranks	368	214.88	79,075.00			
	Ties	239					
11. Downloading music	Negative ranks	45	153.87	6924.00	−16.03[b]	0.000*	0.44
	Positive ranks	405	233.46	94,551.00			
	Ties	200					
12. Listening to music	Negative ranks	123	113.87	14,006.50	−1.616[b]	0.106	0.04
	Positive ranks	128	137.65	17,619.50			
	Ties	399					
13. Downloading film/video	Negative ranks	40	171.21	6848.50	−17.39[b]	0.000*	0.48
	Positive ranks	459	256.87	117,901.50			
	Ties	151					

(continued)

Table 2 (continued)

Purposes	Laptop–mobile phone	N	Mean rank	Sum of ranks	Z	p	r
14. Watching film/video	Negative ranks	31	135.02	4185.50	-17.66^b	0.000*	0.49
	Positive ranks	445	245.71	109,340.50			
	Ties	174					
15. Playing games	Negative ranks	96	148.04	14,212.00	-7.944^b	0.000*	0.22
	Positive ranks	238	175.35	41,733.00			
	Ties	316					
16. Visiting social network (Facebook, Twitter, etc.)	Negative ranks	55	150.20	8261.00	-11.97^b	0.000*	0.33
	Positive ranks	294	179.64	52,814.00			
	Ties	301					
17. Spoken communication	Negative ranks	479	281.18	134,685.00	-17.781^a	0.000*	0.49
	Positive ranks	55	148.36	8160.00			
	Ties	116					
18. Written communication (SMS, e-mail, Twitter, etc.)	Negative ranks	151	149.29	22,543.50	-6.019^b	0.000*	0.17
	Positive ranks	222	212.65	47,207.50			
	Ties	277					
19. Reading newspaper	Negative ranks	44	135.15	5946.50	-17.822^b	0.000*	0.49
	Positive ranks	461	264.25	121,818.50			
	Ties	145					
20. Watching TV	Negative ranks	44	195.92	8620.50	-15.481^b	0.000*	0.44
	Positive ranks	408	229.80	93,757.50			
	Ties	198					
21. Keeping abreast of national and international developments	Negative ranks	36	149.01	5364.50	-18.275^b	0.000*	0.51
	Positive ranks	481	267.23	128,538.50			
	Ties	133					
22. Banking transactions	Negative ranks	54	147.50	7965.00	-12.999^b	0.000*	0.36
	Positive ranks	316	191.99	60,670.00			
	Ties	280					
23. Shopping	Negative ranks	29	112.38	3259.00	-15.441^b	0.000*	0.43
	Positive ranks	348	195.39	67,994.00			
	Ties	273					
24. Surfing the internet	Negative ranks	46	166.76	7671.00	-16.297^b	0.000*	0.45
	Positive ranks	423	242.42	102,544.00			
	Ties	181					
25. Downloading programmes	Negative ranks	35	134.00	4690.00	-17.849^b	0.000*	0.50
	Positive ranks	455	254.08	115605.00			
	Ties	160					
26. Showing-off	Negative ranks	50	74.90	3745.00	-5.971^b	0.000*	0.17
	Positive ranks	125	93.24	11655.00			
	Ties	475					

*$p < 0.05$
[a]Based on positive ranks
[b]Based on negative ranks

$r_{(20)} = 0.44$; $r_{(10)} = 0.43$; $r_{(23)} = 0.43$; $r_{(22)} = 0.36$; $r_{(16)} = 0.33$). That means prospective teachers stated to use laptops for these purposes practically more frequently than mobile phones. However, effect sizes for five remaining items were estimated to be small ($r_{(3)} = 0.22$; $r_{(15)} = 0.22$; $r_{(18)} = 0.17$; $r_{(26)} = 0.17$; $r_{(1)} = 0.14$), indicating poor practical significance.

5.3 Results About Gender

In the present study we also investigated whether participating students' gender caused any significant differences in their frequency of using these mobile tools for different purposes. Results of non-parametric Mann Whitney U tests for each item (see Table 3) revealed that men used mobile phones significantly more frequently than women to present projects/assignments, to prepare for exams, to download film/video, to watch TV, to do shopping, to download programmes and to show-off. On the other hand, women were observed to use mobile phones significantly more

Table 3 Comparison of prospective teachers' frequency of using mobile phones for different purposes by gender

Purpose of mobile phone use	Gender	N	Mean rank	Sum of ranks	U	z	p	r
6. Presenting projects/assignments	Women	430	312.69	134,458	41,793	−2794	0.005*	0.11
	Men	220	350.53	77,117				
8. Preparing for exams	Women	430	314.88	135,397	42,732	−2256	0.024*	0.088
	Men	220	346.26	76,178				
12. Listening to music	Women	430	348.53	149,868	37,397	−4641	0.000*	0.182
	Men	220	280.49	61,707				
13. Downloading film/video	Women	430	310.19	133,383	40,718	−3105	0.002*	0.122
	Men	220	355.42	78,192				
18. Written communication (SMS, e-mail, Twitter, etc.)	Women	430	338.71	145,644	41,621	−2609	0.009*	0.102
	Men	220	299.69	65,931				
19. Reading newspaper	Women	430	314.19	135,102.5	42,437.5	−2287	0.022*	0.09
	Men	220	347.60	76,472.5				
20. Watching TV	Women	430	313.38	134,755.5	42,090.5	−2594	0.009*	0.102
	Men	220	349.18	76,819.5				
23. Shopping	Women	430	312.61	134,421	41,756	−2953	0.003*	0.116
	Men	220	350.70	77,154				
25. Downloading programmes	Women	430	309.13	132,926.5	40,261.5	−3315	0.001*	0.13
	Men	220	357.49	78,648.5				
26. Showing-off	Women	430	312.88	134,540.5	41,875.5	−2790	0.005*	0.109
	Men	220	350.16	77,034.5				

*$p < 0.05$

Table 4 Comparison of prospective teachers' frequency of using laptop for different purposes by gender

Purpose of laptop use	Gender	N	Mean rank	Sum of ranks	U	z	p	r
4. Making research for projects/assignments	Women	430	342.24	147,165	40,100	−3753	0.000*	−0.15
	Men	220	292.77	64,410				
5. Preparing projects/assignments	Women	430	341.75	146,952.50	40,312.5	−3519	0.000*	−0.14
	Men	220	293.74	64,622.50				
11. Downloading music	Women	430	336.73	144,792	42,473	−2235	0.025*	−0.09
	Men	220	303.56	66,783				
12. Listening to music	Women	430	341.84	146,991	40,274	−3298	0.001*	−0.13
	Men	220	293.56	64,584				
14. Watching film/video	Women	430	344.07	147,950.50	39,314.5	−4013	0.000*	−0.16
	Men	220	289.20	63,624.50				
15. Playing games	Women	430	305.79	131,489	38,824	−3839	0.000*	−
	Men	220	364.03	80,086				
18. Written communication (SMS, e-mail, Twitter, etc.)	Women	430	338.67	145,630	41,635	−2646	0.008*	−0.10
	Men	220	299.75	65,945				
22. Banking transactions	Women	430	309.45	133,063	40,398	−3118	0.002*	−0.12
	Men	220	356.87	78,512				
23. Shopping	Women	430	310.55	133,537	40,872	−2915	0.004*	−0.11
	Men	220	354.72	78,038				

*$p < 0.05$

frequently than men to listen to music and for written communication (SMS, e-mail, Twitter, etc.). However, all of these differences represented small-sized effect ($\leq r = 0.182$), indicating practically negligible effect of gender.

As for the prospective teachers' purposes of using laptops, Mann Whitney U tests for each item (see Table 4) revealed that women use laptops significantly more frequently than men to make research for projects/assignments, to prepare projects/assignments, to download music, to listen to music, to watch film/video, and for written communication (SMS, e-mail, Twitter, etc.). Men, on the other hand, stated to use laptops more frequently than women to play games, for banking transactions and for shopping. However, all of these differences represented small-sized effect ($\leq r = 0.15$), indicating practically negligible effect of gender.

5.4 Results About Grade

The Kruskal Wallis analysis of participants' purposes of using mobile phones by grade variable revealed significant differences for only one item: "*12. Listening to music*" [$X^2 = 15.228$, $p < 0.05$]. According to the post hoc Mann Whitney U test results first-year students (*MeanRank* = 162.16, U = 8523, $p < 0.05$, r = 0.13), second-year students (*MeanRank* = 165.47, U = 9774, $p < 0.05$, r = 0.10) and third-year students (*MeanRank* = 173.92, U = 9750, $p < 0.05$, r = 0.13) used mobile phones more frequently to listen to the music than fourth-year students (*MeanRank*$_1$ = 131.38, *MeanRank*$_2$ = 140.31 and *MeanRank*$_3$ = 140.14, respectively). Since the estimated effect sizes or these differences were small ($\leq r$ = 0.13), the effect of grade was considered negligible in practice.

As for the comparison of prospective teachers' purposes of laptop use by grade, Kruskal Wallis analysis yielded a statistically significant difference only for "*12. Listening to music.*" once again [X^2 = 13.739, $p < 0.05$]. The post hoc Mann Whitney U tests revealed that first-year students (*MeanRank* = 161.34, U = 8648.5, $p < 0.05$, r = 0.12) and third-year students (*MeanRank* = 170.62, U = 10,333, $p < 0.05$, r = 0.10) used laptops more frequently to listen to music than fourth-year students (*MeanRank*$_1$ = 132.28 and *MeanRank*$_3$ = 144.31, respectively). Since the estimated effect sizes for these differences were small ($\leq r$ = 0.12), the effect of grade was considered negligible in practice.

5.5 Results About Frequency of Connecting to Internet with Mobile Tools

Connecting to Internet is an important part of mobile learning. Thus, we next analysed participants' frequencies of connecting to Internet with their mobile tools. As it is seen in Table 5, out of 650 students having a mobile phone 642 answered the question about the frequency of connecting to Internet with their mobile phones ranging between 1 point for *never* and 5 points for *always*. It was found that 160 students (24.9 %) *never* connected to Internet via their mobile phones, 111 (17.3 %) connected *rarely,* 126 (19.6 %) connected *sometimes,* 129 (20.1 %) connected *usually,* and 116 (18.1 %) connected *always* ($\bar{x}$ = 2.89, sd = 1.44 and median = 3). This indicates that prospective teachers did not connect to Internet via mobile phone very commonly.

Out of 650 students having a laptop 647 answered the question about the frequency of connecting to Internet with their laptops ranging between 1 point for *never* and 5 points for *always*. It was found that 11 students (1.7 %) *never* connected to Internet via their laptops, 51 (7.9 %) connected *rarely,* 137 (21.2 %) connected *sometimes,* 198 (30.6 %) connected *usually,* and 250 (38.6 %) connected *always,* ($\bar{x}$ = 3.97, sd = 1.03 and median = 4).

Finally, the frequency of connecting to Internet with both mobile tools was compared using Wilcoxon test. The results of the analysis are presented in Table 6. The results revealed that prospective teachers connect to Internet via laptops significantly more frequently than they do so via mobile phones [z = 13.729, $p < 0.05$, r = 0.38 representing *medium* effect size].

Table 5 Frequency of connecting to internet with mobile tools

Frequency of connecting to internet via	Never		Rarely		Sometimes		Usually		Always		Total		$\bar{x}$	s	Med.
	f	%	f	%	f	%	f	%	f	%	f	%			
Mobile phone	160	24.9	111	17.3	126	19.6	129	20.1	116	18.1	642	100	2.89	1.14	3
Laptop	11	1.7	51	7.9	137	21.2	198	30.6	250	38.6	647	100	4.97	1.03	4

Table 6 Comparison of frequency of connecting to internet with both mobile tools

Frequency of connecting to Internet		N	Mean rank	Sum of ranks	Z	p	r
Laptop–mobile phone	Negative ranks	100	155.92	15,592	$-13,729^a$	0.000*	0.38
	Positive ranks	372	258.16	96,036			
	Ties	168					
	Total	640					

*$p < 0.05$
[a]Based on positive ranks

6 Discussion

This study investigated the prospective teachers' frequency of using two ubiquitous mobile devices, laptops and mobile phones, for different purposes including the educational ones. This purpose served to understand the current situation about prospective teachers regarding mobile learning, which is important because when they become teachers they will be either teaching/having feedback about their subject matter using mobile devices like laptops, tablet PCs, smartphones, etc., or teaching their learners the strategies about using mobile devices for learning.

This study mainly revealed that, in general, prospective teachers used laptops more frequently than mobile phones for almost all categories (i.e. education, entertainment, communication/information and others), except for the purpose of *recording lessons in audio format* and *spoken communication,* where the situation is reversed as can be expected, and *recording lessons in video format* and *listening to music,* where there were no significant differences. This higher rate of laptop use for most of the purposes compared to mobile phones can be said to stem from laptops' versatility. Among other portable devices, laptops or tablet PCs have the most complete and functional systems [32] and offer a variety of advantages such as a standard keyboard and full suite of computing tools [77]. On the other hand, if not smartphones, most mobile phones are more limited in capacity with their limited screen size, keyboard accessibility, memory capacity and processing power [23, 43, 58, 59]. Smartphones, however, as combination of phone and computer technology [32, 62, 78] can do what the average laptops can do, or more. Therefore, the superiority of laptops over mobile phones in terms of versatility could be due to the possibility that most participants owned mobile phones with limited capabilities. In the near future, however, this might cease to be a problem. As a matter of fact, distinctions such as phone, PDA, PC, tablet, etc., are gradually becoming more blurred as all devices become more multifunctional and versatile. It is possible that, in the very near future, most people will only use one device, which most likely would be the smartphone [24].

From the mobile learning perspective, it was especially remarkable that, compared to mobile phones, laptops were used more for educational purposes and

especially for project/assignment-related tasks (e.g. *making research for, preparing
and presenting projects–assignments*), sharing lesson content, following developments in the subject field, personal development, and preparing for exams. This
finding seems to echo and support some previous research. For example, Malaysian
students were found to perceive mobile phones less suitable for education [23].
Also, Şad and Göktaş [24] reported that prospective Turkish teachers found laptops
to be potentially stronger than mobile phones as mobile learning tools. Among
these purposes accessing course information (e.g. schedulers, exam results) was
reported to be a common purpose of using mobile devices by the US college
students [29]. Learning by downloading course content and watching real-time
video lectures were reported to be Korean university students' main methods of
mobile learning [30].

   However, in the present study, both laptops and mobile phones, especially the
former, were found to be used less for in-class educational purposes such as taking
notes during lessons and recording lessons in the audio or video formats. Similarly,
Kinash et al. [37] found that taking notes using a mobile tool (iPad) in class was one
of the least common activities. This implies that in-class use of laptops or mobile
phones are not common or welcome. As a matter of fact, mobile learning tools in
general [29] and laptops in particular are notorious for their during-class distraction
factor, negatively affecting understanding of course material and overall course
performance [63].

   As for the mobile phones, they were found to be used mainly for communication
and entertainment purposes with *spoken communication* and *listening to music* at
the top of the list, followed by *written communication (via SMS, e-mail, Twitter,
etc.)* and *visiting social networks (e.g. Facebook, Twitter, etc.).* Similarly,
Economides and Grousopoulou [79] reported that Greek university students use
their mobiles mostly for phone calls and SMS (short message service). Kinash et al.
[37] also found that surfing the web for pleasure or browsing Facebook were the
most common activities among undergraduate students. Educational purposes,
however, were found to be fulfilled by mobile phones less frequently. This finding
concurs with the results from earlier studies. Cui and Wang [71] argued that most
people use cell phones as a communication or recreational tool rather than for
learning or studying. Similarly, Suki and Suki [23] reported that, in the eyes of
students, mobile phones are not for education, but rather beneficial for fun and
communication purposes. Avenoğlu [57] also reported that music and radio are the
most desired mobile phone functions. Franklin [26] cited that college students
consider phones as their single most important form of communication. Actually,
the findings of the previous research are rather controversial concerning the mobile
learning potentials of these tools.

   On the one hand, much research disregards the mobile learning potential of the
mobile phones particularly, due to the reasons stated above and that of mobile
technologies in general by suggesting that they are primarily used for listening to
music [26], surfing the web for pleasure, or browsing Facebook [37]. However,
some other research suggests that students want to continue learning on their
phones [57], are ready for the integration of cell phones in their class work [80], are

positive about using mobile phones, as they are helpful for learning subjects such as biology [81], and enjoy learning new words on their mobile phones [78]. This inconsistency should remind the mobile learning researchers of the principle "What matters is not the mobile tool itself, but how it is integrated with learning and teaching." The failure to obtain expected positive outcomes from a technology should not always be attributed to the technology itself, but to the failure to blend technology and pedagogy [82].

The results regarding gender suggested that though there were statistically significant differences in terms of some purposes, given the lack of practical significance of the differences both male and female prospective teachers can be said to use mobile phones and laptops for various purposes with similar frequencies. The same was also true for the grade variable: all prospective teachers from first to fourth years used mobile phones and laptops for various purposes with similar frequencies in practice. This result implies that four years of education including several ICT-related courses have failed to make a significant difference in prospective teachers' frequency of using the mobile technology especially for educational purposes.

Finally, mobile phones, tablets and netbooks are increasingly used to access the Internet [83]. Among them mobile phones are now the most portable and ubiquitous devices with Internet connectibility [80]. Nevertheless, the present study revealed that, for prospective teachers, connecting to the Internet via mobile phones is not very common and even significantly less common than doing so via laptops.

7 Conclusions and Implications

In this study, we tried to understand prospective teachers' purposes of using laptops and mobile phones from a comparative perspective. Results showed that, compared to mobile phones, laptops are used more frequently for various purposes, especially educational ones. Still, the in-class use of both laptops and mobile phones for educational purposes is not common. Mobile phones were used less for educational purposes, but more for communication and entertainment.

The findings in general suggest a need to promote the mobile learning potential of mobile phones in general and in-class use of laptops in particular. This implies a need to increase awareness about mobile learning among students first. As Terras and Ramsay [83] posit one important challenge facing effective mobile learning is a potential failure to develop metacognitive strategies. They suggest that mobile learners "become aware of how they learn in general and be sensitive in particular to the increased demands of mobile learning and how they can best be managed if mobile learning is to become a reality" (p. 825). Actually, it has become an urgent need for teachers in Turkey to learn how to manage mobile learning, since students around Turkey are now being granted tablet computers by the ministry as a part of a government-supported project called FATIH-Enhancing Opportunities and Improving Technology [84]. This development means that prospective teachers should become ready to manage mobile learning with these tablet computers or similar mobile tools as soon as possible.

The need to raise awareness among prospective teachers, in turn, requires some awareness about the importance of the faculty members' role in adapting mobile learning [29] in order for the integration of mobile learning to be successful. As Corbeil and Valdes-Corbeil [32] report, though most (94 %) students are ready for mobile learning, less of the faculty members (60 %) are ready, or even resistant to mobile learning [26]. Moreover, faculty in general may fail to provide good models for pre-service teachers in terms of shaping their visions about future teaching [85].

As a matter of fact, mobile learning integration requires not only possessing the device, but also the synchronous coordination of infrastructure, technical support and faculty development [26], as well as know-how on the part of instructors [51]. Thus, the faculty members should adapt their roles as "mobile coaches" [34] who "scaffold the [mobile] learning assists according to learner's ability and learning progress in their learning activities…[and]…is expected to monitor learners' needs and provide them with appropriate aid in the [mobile] learning activity" (p. 7).

Therefore, it can be said that growing awareness and developing metacognitive strategies about mobile learning among prospective teachers requires the teacher educators and the administration to take some action. As an example, some communications (information, announcements, etc.) can be sent by the faculty members or dean's office using mobile phones or laptops to increase students' awareness about mobile learning. Faculty members can share their lesson content through mobile learning-friendly devices. Students can be granted tablets, laptops or other devices under scholarship programmes and the infrastructure can be developed to provide students with free wireless Internet throughout the campus. The findings of the present research also suggest a need to investigate, for future researchers, the pros and cons of using laptops and mobile phones in class as learning or teaching tools.

Moreover, the present research was limited to mobile phones in general and laptops, since they were the mobile tools commonly used in the studied context. However, today smartphones should be distinguished from ordinary mobile phones and studied as its own for their educational potentials. Also new trends such as wearables can be studied within the framework of mobile learning.

Acknowledgment We would like to thank our colleague İlhami Bayrak for his help during the study.

References

1. Mishra, P., Koehler, M.J.: Technological pedagogical content knowledge: a framework for teacher knowledge. Teachers Coll. Rec. **108**(6), 1017–1054 (2006)
2. Cuhadar, C., Bulbul, T., Ilgaz, G.: Exploring of the relationship between individual innovativeness and techno-pedagogical education competencies of pre-service teachers. Elem Educ Online **12**(3), 797–807 (2013)

3. Meric, G.: Determining science teacher candidates' self-reliance levels with regard to their technological pedagogical content knowledge. J Theory Practice Educ **10**(2), 352–367 (2014)
4. Ozturk, E., Horzum, M.B.: Adaptation of technological pedagogical content knowledge scale to Turkish. Ahi Evran Üniversitesi Eğitim Fakültesi Dergisi. **12**(3), 255–278 (2011)
5. Şad, S.N., Nalçacı, Öİ.: Öğretmen Adaylarının Eğitimde Bilgi ve İletişim Teknolojilerini Kullanmaya İlişkin Yeterlilik Algıları. Mersin Üniversitesi Eğitim Fakültesi Dergisi. **11**(1), 177–197 (2015)
6. Yavuz-Konokman, G., Yanpar-Yelken, T., Sancar-Tokmak, H.: An investigation of primary school pre-service teachers' perception on their Tpack in terms of a variety factors: Mersin university case. Kastamonu Eğitim Dergisi. **21**(2), 665–684 (2013)
7. Alayyar, G., Fisser, P., Voogt, J.: Developing technological pedagogical content knowledge in pre-service science teachers: support from blended learning. Australas J. Educ. Technol. **28**(8), 1298–1316 (2012)
8. Graham, C., Burgoyne, N., Cantrell, P., Smith, L., Clair, St L., Harris, R.: TPACK development in science teaching: measuring the TPACK confidence of inservice science teachers. Techtrends **53**(5), 70–79 (2009)
9. Şad, S.N., Açıkgül, K., Delican, K.: Eğitim Fakültesi Son Sınıf Öğrencilerinin Teknolojik Pedagojik Alan Bilgilerine (TPAB) İlişkin Yeterlilik Algıları. Kuramsal Eğitimbilim Dergisi [J. Theor. Educ. Sci.] **8**(2), 204–235 (2015)
10. Timur, B., Taşar, M.F.: In-service science teachers' technological pedagogical content knowledge confidences and views about technology-rich environments. CEPS J. **1**(4), 11–25 (2011)
11. Yurdakul Kabakci, I., Odabasi, H.F., Kilicer, K., Coklar, A.N., Birinci, G., Kurt, A.A.: The development, validity and reliability of TPACK-deep: a technological pedagogical content knowledge scale. Comput. Educ. **58**(3), 964–977 (2012)
12. Koehler, M.J., Mishra, P.: What is technological pedagogical content knowledge (TPCK): What is TPCK? In: AACTE Committee on Innovation and Technology (ed.) Handbook of Technological Pedagogical Content Knowledge (TPCK) for Educators, pp. 1–29. New York: Routledge Taylor & Francis (2008)
13. Akbulut, Y., Odabaşı, H., Kuzu, A.: Perceptions of preservice teachers regarding the integration of information and communication technologies in turkish education faculties. Turkish Online J. Educ. Technol. **10**(3), 175–184 (2011)
14. Hirca, N., Simsek, H.: Enhancing and evaluating prospective teachers' technopedagogical knowledge integration towards science subject. Necatibey Fac. Educ. Electron. J. Sci. Math. Educ. **7**(1), 57–82 (2013)
15. Ulas, H., Ozan, C.: The qualification level of primary school teachers' use of educational technology. Atatürk Üniversitesi Sosyal Bilimler Enstitüsü Dergisi. **14**(1), 63–84 (2010)
16. Yılmaz, M.: Instructional technology in training primary school teacher. Gazi Eğitim Fakültesi Dergisi. **27**(1), 155–167 (2007)
17. Tezci, E.: Turkish primary school teachers' perceptions of school culture regarding ict integration. Educ. Tech. Res. Dev. **59**(3), 429–443 (2011)
18. Russell, M., Bebell, D., O'Dwyer, L., ve O'Connor, K.: Examining teacher technology use: implications for preservice and inservice teacher preparation. J. Teacher Educ. **54**, 297–310 (2003)
19. Pereira, O.E., Rodrigues, J.C.: Survey and analysis of current mobile learning applications and technologies. ACM Comput. Surv. **46**(2), 27:1–27:35 (2013)
20. Ebner, M., Nagler, W., Schön, M.: "Architecture students hate twitter and love dropbox" or does the field of study correlates with web 2.0 behavior? In: Proceedings of World Conference on Educational Multimedia, Hypermedia and Telecommunications, pp. 43–53. AACE, Chesapeak VA (2013)
21. Ebner, M., Nagler, W., Schön, M.: Do you mind NSA affair? Does the global surveillance disclosure impact our students? In: Proceedings of World Conference on Educational Multimedia, Hypermedia and Telecommunications, pp. 2307–2312. AACE, Chesapeake VA (2014)

22. Kearney, M., Schuck, S., Burden, K., Aubusson, P.: Viewing mobile learning from a pedagogical perspective. Res. Learn. Technol. **20**(1) (2012)
23. Suki, N.M., Suki, N.M.: Using mobile device for learning: from students' perspective. US-China Educ. Rev. **A1**, 44–53 (2011)
24. Şad, S.N., Göktaş, Ö.: Preservice teachers' perceptions about using mobile phones and laptops in education as mobile learning tools. Br. J. Educ. Technol. BJET **45**(4), 606–618 (2014)
25. El-Hussein, M.O.M., Cronje, J.C.: Defining mobile learning in the higher education landscape. Educ. Technol. Soc. **13**(3), 12–21 (2010)
26. Franklin, T.: Mobile learning: at the tipping point. Turk. Online J. Educ. Technol. TOJET **10** (4), 261–276 (2011)
27. Kalinic, Z., Arsovski, S., Stefanovic, M., Arsovski, Z., Rankovic, V.: The development of a mobile learning application as support for a blended e-learning environment. Tech. Technol. Educ. Manag. **6**(4), 1345–1355 (2011)
28. Isik, A.H., Ozkaraca, O., Guler, I.: Mobil Öğrenme ve Podcast. In: Akademik Bilişim'11, İnönü University Turkey (2011)
29. Cheon, J., Lee, S., Crooks, S.M., Song, J.: An investigation of mobile learning readiness in higher education based on the theory of planned behavior. Comput. Educ. **59**, 1054–1064 (2012)
30. Park, S.Y., Nam, M.W., Cha, S.B.: University students' behavioral intention to use mobile learning: evaluating the technology acceptance model. Br. J. Educ. Technol. **43**(4), 592–605 (2012)
31. Ebner, M., Nagler, W., Schön, M.: Have they changed? Five years of survey on academic net-generation. In: Proceedings of World Conference on Educational Multimedia, Hypermedia and Telecommunications 2012, pp. 343–353. AACE, Chesapeake, VA (2012)
32. Corbeil, J.R., Valdes-Corbeil, M.E.: Are you ready for mobile learning? Educ. Q. **30**(2), 51–58 (2007)
33. Seppälä, P., Alamäki, H.: Mobile learning in teacher training. J. Comput. Assist. Learn. **19**, 330–335 (2003)
34. Jeng, Y.L., Wu, T.T., Huang, Y.M., Tan, Q., Yang, S.J.H.: The add-on impact of mobile applications in learning strategies: a review study. Educ. Technol. Soc. **13**(3), 3–11 (2010)
35. Peng, H., Su, Y.J., Chou, C., Tsai, C.C.: Ubiquitous knowledge construction: mobile learning re-defined and a conceptual framework. Innov. Educ. Teaching Int. **46**(2), 171–183 (2009)
36. Keskin, N., Metcalf, D.: The current perspectives, theories and practices of mobile learning. Turk. Online J. Educ. Technol. TOJET **10**(2), 202–208 (2011)
37. Kinash, S., Brand, J., Mathew, T.: Challenging mobile learning discourse through research: student perceptions of blackboard mobile learn and iPads. Australas. J. Educ. Technol. **28**(4), 639–655 (2012)
38. Wang, M., Shen, R.: Message design for mobile learning: learning theories, human cognition and design principles. Br. J. Educ. Technol. **43**(4), 561–575 (2012)
39. Wu, W., Wu, Y.J., Chen, C., Kao, H., Lin, C., Huang, S.: Review of trends from mobile learning studies: a meta-analysis. Comput. Educ. **59**(2), 817–827 (2012)
40. Chen, I.-J., Chang, C.-C., Yen, J.-C.: Effects of presentation mode on mobile language learning: a performance efficiency perspective. Australas. J. Educ. Technol. **28**(1), 122–137 (2012)
41. Georgieva, E., Smrikarov, A., Georgiev, T.: A general classification of mobile learning systems. In: International Conference on Computer Systems and Technologies—CompSysTech' 2005 (2005)
42. Shih, J.L., Chuang, C.W., Hwang, G.J.: An inquiry-based mobile learning approach to enhancing social science learning effectiveness. Educ. Technol. Soc. **13**(4), 50–62 (2010)
43. Motiwalla, L.F.: Mobile learning: a framework and evaluation. Comput. Educ. **49**, 581–596 (2007)
44. Grimus, M., Ebner, M.: M-learning in sub Saharan Africa context—what is it about. In: Proceedings of World Conference on Educational Multimedia, Hypermedia and Telecommunications, pp. 2028–2033. AACE, Chesapeake, VA (2013)

45. Gülsecen, S., Gursul, F., Bayrakdar, B., Cilengir, S., Canim, S.: New generation mobil learning tool: podcast. In: Proceedings of XII. Academic Information Conference, Muğla University Turkey (2010)
46. Houser, C., Thornton, P., Kluge, D.: Mobile learning: cell phones and PDAs for education. In: Proceedings of the International Conference on Computers in Education (ICCE'02) (2002)
47. Nordin, N., Embi, M.A., Yunus, MM.: Mobile learning framework for lifelong learning. Procedia Soc. Behav. Sci. 7(C), 130–138 (2010)
48. Saran, M., Seferoglu, G.: Supporting foreign language vocabulary learning through multimedia messages via mobile phones. Hacettepe Univ. J. Educ. 38, 252–266 (2010)
49. Uzunboylu, H., Ozdamli, F.: Teacher perception for M-learning: scale development and teachers' perceptions. J. Comput. Assist. Learn. 27, 544–556 (2011)
50. Kim, J.Y.Y.: A survey on mobile-assisted language learning. Mod. Engl. Educ. 7(2), 57–69 (2006)
51. Menkhoff, T., Bengtsson, M.L.: Engaging students in higher education through mobile learning: lessons learnt in a chinese entrepreneurship. Educ. Res. Policy Prac. 1, 225–242 (2012)
52. Sølvberg, A.M., Rismark, M.: Learning spaces in mobile learning environments. Act. Learn. High Educ. 13(1), 23–33 (2012)
53. Şad, S.N.: Using mobile phone technology in EFL classes. Engl. Teaching Forum 46(4), 34–39 (2008)
54. Şad, S.N., Akdağ, M.: İngilizce Dersinde Cep Telefonlarıyla Üretilen Sözlü Performans Ödevlerinin Yazılı Performans Ödevleriyle Karşılaştırılması. Türk Eğitim Bilimleri Dergisi. 8(3), 719–740 (2010)
55. Arslan, İ.: Developing reusable mobile learning objects using flash. In: 5th International Computer & Instructional Technologies Symposium, Fırat University Elazığ (2011)
56. Yang, S.: Exploring college students' attitudes and self-efficacy of mobile learning. Turk. Online J. Educ. Technol. TOJET 11(4), 148–154 (2012)
57. Avenoglu, B.: Using mobile communication tools in web based instruction. Unpublished master thesis, Middle East Technical University, Ankara Turkey (2005)
58. Huber, S., Ebner, M.: iPad human interface guidelines for M-learning. In: Berge, Z.L., Muilenburg, L.Y. (eds.) Handbook of Mobile Learning, pp. 318–328. Routledge, New York (2013)
59. Bal, Y., Arici, N.: Mobile based learning materials preparation. Bilişim Teknolojileri Dergisi. 4(1), 7–12 (2011)
60. Oran, M.K., Karadeniz, S.: Mobile Learning Role in Internet Based Distance Education. Dumlupınar University Turkey, Akademik Bilişim Konferansı (2007)
61. Riad, A.M., El-Ghareeb, H.A.: A service oriented architecture to integrate mobile. Turkish Online J. Distance Educ. TOJDE 9(2), 200–219 (2008)
62. Yılmaz, Y.: Investigating the awareness levels of postgraduate students and academics towards mobile learning. Unpublished master thesis Dokuz Eylül University, İzmir Turkey (2011)
63. Fried, B.C.: In-class Laptop use and its effects on student learning. Comput. Educ. 50, 906–914 (2008)
64. Malita, L., Martin, C.: Digital storytelling as web passport to success in the 21st century. Procedia Soc. Behav. Sci. 2, 3060–3064 (2010)
65. Keskin, N.O.: Mobile learning technologies and tools. In: Proceedings of XII. Academic Information Conference, Muğla University Turkey (2010)
66. Maria, V., Eythimios, A.: Mobile educational features in authoring tools for personalised tutoring. Comput. Educ. 44, 53–68 (2005)
67. Hangul, E., Kalayci, T.E., & Ugur, A.: Üç Boyutlu Grafik Teknolojilerinin Mobil Öğrenme Alanı Ile Bütünleştirilmesi. In: 2nd International Future-Learning Conference, İstanbul University Istanbul (2008)
68. Oberg, A., Daniels, P.: Analysis of the effect a student-centred mobile learning instructional method has on language acquisition. Comput. Assis. Lang. Learn. 26(2), 177–196 (2013)

69. Gromik, A.N.: Cell phone video recording feature as a language learning tool: a case study. Comput. Educ. **58**, 223–230 (2012)
70. Ebner, M.: The influence of twitter on the academic environment. In: Patrut, B., Patrut, M., Cmeciu, C. (eds.) Social Media and the New Academic Environment: Pedagogical Challenges, pp. 293–307. IGI Global, Hershey (2013)
71. Cui, G., Wang, S.: Adopting cell phones in EFL teaching and learning. J. Educ. Technol. Dev. Exch. **1**, 69–80 (2008)
72. Park, Y.: A pedagogical framework for mobile learning: categorizing educational applications of mobile technologies into four types. Int. Rev. Res. Open Dist. Learn. **12**(2), 78–102 (2011)
73. Bulun, M., Gulnar, B., Guran, M.S.: Mobile technologies in education. Turk. Online J. Educ. Technol. **3**(2), 165–169 (2004)
74. Saran, M., Seferoglu, G., Cagiltay, K.: Mobile assisted language learning: English pronunciation at learners' fingertips. Eurasian J. Educ. Res. **34**, 97–114 (2009)
75. Fraenkel, J.R., Wallen, N.E., Hyun, H.H.: How to design and evaluate research in education, 8th edn. McGraw-Hill, New York (2012)
76. Field, A.: Discovering Statistics Using Spss, 3rd edn. Sage Publications, London (2009)
77. Wong, W.: Tools of the trade: how mobile learning devices are changing the face of higher education. Community Coll. J. **82**(5), 54–61 (2012)
78. Cavus, N., Ibrahim, D.: M-learning: an experiment in using Sms to support learning new english language words. Br. J. Educ. Technol. **40**(1), 78–91 (2009)
79. Economides, A.A., Grousopoulou, A.: Use of mobile phones by male and female Greek students. Int. J. Mobile Commun. (IJMC) **6**(6), 729–749 (2008)
80. Van De Bogart, W.: Adopting cell phones in the classroom: a study of students' attitudes and behaviors on using cell phones both in and out of the classroom. In: Proceedings of the International Conference on Intellectual Capital, Knowledge Management & Organizational Learning, pp. 571–579 (2011)
81. Gedik, N., Hanci-Karademirci, A., Kursun, E., Cagiltay, K.: Key instructional design issues in a cellular phone-based mobile learning project. Comput. Educ. **58**(4), 1149–1159 (2012)
82. Şad, S.N., Özhan, U.: Honeymoon with Iwbs: a qualitative insight in primary students' views on instruction with interactive whiteboard. Comput. Educ. **59**(4), 1184–1191 (2012)
83. Terras, M.M., Ramsay, J.: The five central psychological challenges facing effective mobile learning. Br. J. Educ. Technol. **43**(5), 820–832 (2012)
84. MoNE (2014) Eğitimde FATIH projesi. (FATIH project in education). http://fatihprojesi.meb.gov.tr/site/
85. Gürbüztürk, O., Duruhan, K., Şad, S.N.: The association between preservice teachers' previous formal education experiences and their visions about their future teaching. Elem. Educ. Online **8**(3), 923–934 (2009)

Flexible and Contextualized Cloud Applications for Mobile Learning Scenarios

Alisa Sotsenko, Janosch Zbick, Marc Jansen and Marcelo Milrad

Abstract This chapter describes our research efforts related to the design of mobile learning (m-learning) applications in cloud-computing (CC) environments. Many cloud-based services can be used/integrated in m-learning scenarios, hence, there is a rich source of applications that could easily be applied to design and deploy those within the context of cloud-based services. Here, we present two cloud-based approaches—a flexible framework for an easy generation and deployment of mobile learning applications for teachers, and a flexible contextualization service to support personalized learning environment for mobile learners. The framework provides a flexible approach that supports teachers in designing mobile applications and automatically deploys those in order to allow teachers to create their own m-learning activities supported by mobile devices. The contextualization service is proposed to improve the content delivery of learning objects (LOs). This service allows adapting the learning content and the mobile user interface (UI) to the current context of the user. Together, this leads to a powerful and flexible framework for the provisioning of potentially ad hoc mobile learning scenarios. We provide a description of the design and implementation of two proposed cloud-based approaches together with scenario examples. Furthermore, we discuss the benefits of using flexible and contextualized cloud applications in mobile learning scenarios. Hereby, we contribute to this growing field of research by

A. Sotsenko (✉) · J. Zbick · M. Jansen · M. Milrad
LNU: Linnaeus University, 351 95 Växjö, Sweden
e-mail: alisa.sotsenko@lnu.se

J. Zbick
e-mail: janosch.zbick@lnu.se

M. Jansen
e-mail: marcbjorn.jansen@lnu.se; marc.jansen@hs-ruhrwest.de

M. Milrad
e-mail: marcelo.milrad@lnu.se

M. Jansen
University of Applied Sciences Ruhr West, Dümptener Str. 45,
45476 Mülheim an der Ruhr, Germany

© Springer International Publishing Switzerland 2016
A. Peña-Ayala (ed.), *Mobile, Ubiquitous, and Pervasive Learning*,
Advances in Intelligent Systems and Computing 406,
DOI 10.1007/978-3-319-26518-6_7

167

exploring new ways for designing and using flexible and contextualized cloud-based applications that support m-learning.

Keywords Mobile learning · Contextualization · Contextualized service · Cloud computing · Cloud-based services · Context modeling

Abbreviations

CC	Cloud computing
GPS	Global positioning system
ICT	Information and communication technologies
LO	Learning object
m-learning	Mobile learning
MVSM	Multi-dimensional vector space model
RCM	Rich context model
TEL	Technology enhanced learning
UI	User interface

1   Introduction

The use and integration of information and communication technologies (ICT) in the field of education is constantly increasing. In classrooms, a trend of a transition from traditional teaching methods to digital supported education can be identified [1]. There are some indications that the introduction of the latest ICT developments in classroom settings can improve the quality of teaching and learning [2].

Recent developments in the field of technology enhanced learning (TEL) have led to a renewed interest in new learning approaches and technologies that can be widely used and adapted for different forms of learning. Currently, mobile devices are used in everyday life activities. m-Learning has been defined as the process of learning with the use of mobile devices to access the educational resources, services from any place, and any time where the learner takes advantages of the learning opportunities offered by mobile technologies [3]. Mobile technologies can facilitate learning outside the classroom in order to enhance the learning experience [4]. Also, learning materials are no longer limited to traditional materials like books [5].

A challenge that arises with the growing role of mobile technologies in the field of education is the importance to provide end users, in particular teachers, with the possibility to author and deploy their own mobile applications [6] as they usually do not have the technical and programming skills to develop mobile applications suiting their requirements. So, one area of concern that developers and researchers are exploring is how to give end users the possibility to create and author their mobile applications.

M-learning activities offer also learners opportunities to independently explore processes that involve the gain of knowledge and own experience. Here, the context of the learner plays an important role in supporting the interactions between mobile devices and the environment in which the learning is taking place. By using contextual information, learners can be supported or advised via a mobile application in order to help them to find a solution for the given problem in the real world. Therefore, new ways of interpretation and the consideration of contextual information of mobile learners are necessary.

Currently, cloud-computing (CC) solutions are often used to overcome some of the limitations of mobile devices, desktop computers, or server systems, especially to improve accessibility and interoperability [7]. In this chapter, we explore and present how novel uses of CC can contribute with some advancement to the field of TEL. CC increases the flexibility of modern applications while at the same time improving security aspects, such as availability, data storage, or communication. Furthermore, one major aspect in CC solutions is the accessibility of the provided services through a standardization of interfaces.

With respect to learning scenarios, a different perspective to these abstracted features of CC services (referred to in this chapter as cloud services) provides new ways to conceptualize and deploy emerging services such as contextualization and flexible authoring tools.

Flexible and contextualized mobile applications can be used as entry points for value-adding functions both in formal and informal learning settings, remote and colocated situations and in synchronous or asynchronous scenarios. Using cloud-based services allows increasing flexibility, availability, and the accessibility of services through standardized methods, and additionally it allows the usage of off-the-shelf software, which saves implementation efforts and development time. Informal learning scenarios can particularly benefit from this approach, as it allows for an easier contextualization of the learning experience, which is still a hot topic, in the domain of m-learning.

In this chapter, we address how the field of TEL has been affected by CC technologies by reviewing several studies and applications. The benefits of using cloud-based services for the development of new personalized m-learning applications and services are described. In Sect. 2, we provide arguments about using CC technologies for developing flexible and personalized m-learning applications and services. Section 3, first presents and discusses relevant efforts connected to the use of CC for educational purposes, and already existing cloud-based services for supporting the development of m-learning applications for students and teachers. Second, we provide a few examples on how to support the contextualization of learners by using mobile devices and CC capabilities. The following Sect. 4 describes our cloud-based solutions is contextualized service. Section 4 follows by describing the proposed services provided by a contextualized service with several examples on how they can be used for developing m-learning applications. In Sect. 5 we describe two scenario examples in order to investigate the advantages and benefits of using cloud-based services for suggested contextualized approach. Section 6 concludes the chapter by providing a summary and a discussion of ideas for future directions of research and development.

2 Motivation

Currently, CC solutions have been focused more on the using CC capabilities to create a learning environment for a specific lab, course, assignment, or lesson [8] rather than adding new possible capabilities by using existing ones [7]. For instance, using the CC power for processing vast amount of data about the learner's situation, scalability for supporting a large amount of students, flexibility in using different available services—all these together provide new opportunities to create a learning environment that will be adapted to the current learner's needs and situation.

These examples are not only taking the advantages of the ubiquity of the CC to support m-learning but also shows additional functions and services that can be implemented and used to improve TEL activities [9].

One of the main advantages of using CC technologies together with mobile devices is to enhance the computational capabilities of these resource-constrained units in order to provide rich user experiences [10]. Therefore, the enrichment of the learner's experiences and activities it demands the development of personalized CC services.

2.1 Cloud-Based Services for Teaching

There are a number of cloud-based services that provide opportunities for teachers to create virtual desktop environments with preconfigured software and learning resources [8]. Those services provide an access to different software applications (e.g., Scilab, R) in order to flexibly organize a learning environment for any group of students. However, there are no cloud-based services that can provide opportunities for teachers to design their own learning scenarios (e.g., field trips) and deploy them in a cloud as an m-learning application and finally distribute it among students. Moreover, not all teachers have knowledge about how to configure cloud environments for a certain learning environment. We think that cloud-based solutions should be more flexible in terms of easy and fast configuration of learning scenarios for teachers, and to deploy them as mobile applications for students.

There are many cloud-based services that provide ubiquitous computer power, processing, and storage capabilities together with different software applications to support learning environments suitable to a certain student's learning task [11]. However, few research efforts have been carried out toward supporting an adaptation of cloud-based learning services to the current learner's needs and situation in order to improve the learning performance. Furthermore, it is desirable to support the convenient format representation of learning materials, to define learning problems/difficulties and to evaluate the learning progress. We think that the most salient CC features (e.g., scalability, high availability, flexibility) provide opportunities to create highly personalized services that can be beneficially used both for teachers and learners.

2.2 Benefits of Using Cloud-Based Services

In this subsection, we describe some of the benefits for learners and teachers with regard to mobile applications running in a CC environment:

- Learners can run different software applications on their mobile devices anytime and anywhere. Teachers can provide different software applications available in a cloud environment without additional installation efforts and cost. Additionally, teachers can test a variety of apps from different providers to find out which ones are best for them.
- Teachers can create learning repositories for sharing learning resources among students; learners have access to these resources anytime and from any device [10]. Additionally, it saves the cost of learning materials [12]. Since, students and teachers can share learning e-books, video tutorials with each other in a cloud.
- Using CC technologies can increase learners' engagement and interactions with m-learning applications by enhancing features and functionalities of mobile devices [13].
- Students might use different portable devices during the learning process therefore m-learning applications should be accessible on all of them. Cloud-based services provide the possibility to run learning applications on multiple devices (e.g., tablets, mobile devices, etc.), as long as the device has an internet access [10, 13].
- Scalable storage capacity and processing power allow developing m-learning applications to support large numbers of students [2, 14–16].

One of the possible drawbacks of using CC in the field of TEL is that it requires specific technological skills to configure different learning environments for a lesson, study, and lab that not all teachers may have [17].

The suggested cloud-based solutions we are proposing, namely the web-based framework and the contextualized service pose new challenges. In the case of the web-based framework, the system should be used by a significant amount of users. Since no programmer needs to be involved in the process of generating mobile applications, the potential of the existence of a massive amount of applications is given. One potential solution to address this issue is the deployment of the components, in this case, the mobile application, in a cloud-based system. As mentioned before, cloud-based systems provide benefits with respect to scalability and reliability.

In the case of information contextualization, the processing of huge amount of contextual data requires additional computational resources, storage capacity, and flexible algorithms for analyzing these data sets. A cloud-based solution allows providing richer/more detailed descriptions of the user's context by using additional cloud-based services together with the power of CC (e.g., Amazon Elastic Map and Reduce service). Furthermore, it allows using contextual data from different m-learning scenarios for providing recommendations based on historical data. To address the challenges mentioned above the following research questions have been formulated and they serve as the foundations that guide our efforts:

- *How can a cloud-based solution improve the contextualization approach for mobile users in learning scenarios?*
- *Which cloud-based services can be used for supporting a contextualization approach for mobile users in learning scenarios?*

In the next section, we present an overview of related research efforts in this domain.

3   Related Work

Traditional m-learning applications have limited access to learning resources [13, 18], limited offline data usage support, data sharing, and social-technical issues for teamwork [19]. Cloud-based learning applications can help to solve/overcome these limitations. For example, a rich mobile multimedia cloud-based service enables access rich multimedia content from any mobile device or platform [15].

Using a service-oriented system, "Teamwork as a Service" [19], allows improving and facilitating social collaboration and learning activities for learners' team, and the Microsoft Mobile Apps provides possibility to use offline data when there are network issues. Additionally, the applications provide much richer availability of services in terms of data size, faster processing speed, and saving battery life [13].

Context-aware m-learning applications require processing a vast amount of contextual data as well as to store this contextual data for further processing. Therefore, context-aware m-learning applications can benefit of using CC technologies [12, 14]. For example, the Amazon Elastic storage can be used for collecting and storing sensor data gathered from mobile devices while Microsoft Azure Machine Learning Service can be used for processing and analyzing a historical contextual data in order to make recommendations to the learner and to support him/her in different context situations.

One of the main advantages of CC capabilities is in supporting context-aware learning activities for both individuals and in groups of learners. Here, the context data should be collected and processed from several learners' mobile devices to provide a real-time feedback to the group of learners. Such m-learning applications can use cloud-based services to monitor, e.g., a mood in the group, the performance of the group, the communication, and learning flow in the group while they performing of a certain learning activity, how students interact with a learning material [20] by using, e.g., educational and learning analytics service [21]. Current context-aware m-learning applications do not have enough computation power and resources to support context-aware learning activities for groups of learners.

In order to investigate in more details which existing cloud-based services and applications can be used in m-learning domain, several studies together with cloud service providers have been reviewed and they are described in the following subsection.

3.1 Classification of Cloud-Computing Services and Applications

CC offers different available services, software applications, and resources for processing a huge amount of data, reducing cost, and increasing flexibility and mobility of information [13]. The widely used technologies, such as social networks together with mobile sensors data, CC, and Internet connectivity makes it possible to provide personalized learning through mobile device [10]. Up to now, several cloud-based services have been suggested to adapt the delivery of learning objects (LOs) on mobile devices including text-to-speech solution for learners in the move [22], support of geo-collaboration for situated learning activities [13, 23].

The work carried out by [23] uses combinations of different cloud services to support new forms of TEL activities with adaptation to the learner's style. The current solutions are based on using learning analytics techniques [24] for processing students' learning activities to predict learner's performance or issues by, e.g., monitoring the logs in a cloud. Other challenges described in [10] address problems related to learners that can combine various applications on their mobile devices (e.g., calendar, editor, notes) to configure a personalized cloud learning environment that utilize content and services available from the cloud for their individual needs.

Based on the cloud services classification provided in [7] we describe below already existing cloud-based services and applications and how they can be used in m-learning scenarios.

Cloud-based Communication Services. They are utilized for supporting learner–teacher and learner–learner interactions on a remote or colocated mode. Learners use different types of collaboration, and therefore, use different applications or tools to satisfy their needs and goals. For example, for group/team collaboration the most used technologies either social networks applications (e.g., Facebook Chat, Twitter) or chat-based applications (e.g., Skype, Viber, WeChat) for interaction between their group/team members. For asking questions about a certain problem/issue or answering with a solution for the given problem/issue usually a question and answer sites (e.g., stackoverflow.com, mathoverflow.net) are used by learners, and for collaborative paper working a set of tools (e.g., the Google Apps for Education Suite [25]) can be used to simultaneously communicate and work in the team/group.

For learner–teacher communication learners usually prefer to use email services (e.g., Mailbox, CloudMagic Email) or learning platforms' forums (e.g., Moodle forum). In case of m-learning scenario the mentioned above communication services can be used either through cloud-based mobile applications (e.g., ZOOM Cloud Meetings, the Google Apps for Education Suite is available for tablets and phones) or cloud-based mobile services (e.g., Push notifications). The efforts carried out by [26] show that cloud-based communication services can help teachers to know the current learner situation and to improve the communication between them.

Cloud-based Repository Services. They are used to store, share, and retrieve learning materials or resources in the cloud. The most popular examples of such services, just to mention some of them, are Dropbox, OneDrive, Box, Amazon Cloud Drive, Google Drive that available on iPhone and Android mobile platforms. Learners can use these tools to perform different tasks. For example, Dropbox is used for sharing, accessing different types of files between other learners, while the Box application supports additionally a group work, including assigning tasks and tracking file versions for each team member. This example can be used, e.g., to evaluate the contribution and the performance either of the group of learners itself or individual learner in the group. Such services allow accessing the large number of LOs via mobile through Internet at anywhere and anytime.

Hence, depending on the file size of LO and Internet connection it might take some time to get it. Therefore, such services as DropBox, OneDrive, Google Drive offers offline accessing and viewing files on the mobile device due to Internet connectivity issues. But those files that should be accessed in offline mode should be specified in the application in advanced and in online mode. This offline feature allows for learners learning in anywhere (e.g., sitting in the train, in the park) and anytime with their mobile devices. The added value for teachers is to share the learning resources among a large number of students/learners; tracking both the learner's individual contribution, task's responsibility, performance, and the group work of learners itself. The added value for learners is to accessing the learning resources and sharing their own resources, materials, and works between other learners.

Cloud-based Single-Specialized Services. They are utilized for learning or working on a task that is related to a specific application domain. For example, for learning supported by audio or video stream processing or for playing and creating digital content anywhere and anytime with a mobile device. Two of the most known single-specialized services are the AutoCad 360, which offers viewing, editing, and sharing AutoCAD files via a mobile device.

Other examples are the Adobe Slate, Premiere Clip, and Voice Services, which allow learners turn any document into a visual storytelling that can be used in museums or field trips. In addition to sharing, and editing a video/audio files, the CyberLink's Mobile App Zone provides the possibility to take a picture of live presentation lecture slides and turn them into PDF files on the mobile devices. An additional example is the Quick Graph application for visualizing plots with high quality 2D and 3D mathematical expressions.

Cloud-based Processing Services. They allow analyzing and processing big data sets with different processing algorithms and methods. Such services allow to support learners in real-time during the performing of a certain learning activity through monitoring learners' interactions with a learning environment and a mobile device. Another example is to analyze and process the log files after a learning activity is finished in order to investigate and understand the workflow and its outcomes by using, e.g., learning analytics services.

This feature enables teachers to analyze the weak and strong aspects of a certain learning activity and offers the possibility to improve it. This could also lead that students can get real-time feedback and support that helps them to successfully perform learning activities. Most of the above described cloud-based services provide an API that can be used as an additional service or combination of services to develop novel custom cloud-based applications for m-learning scenarios.

3.2 Mobile Cloud-Computing Services and Applications

Mobile CC is the combination of mobile application, CC, and Internet connectivity aiming to enhance computational and interactional capabilities of mobile devices toward rich user experience [10]. Many cloud providers offer a huge variety of services for mobile devices called as "Mobile back end as a Service" (e.g., Microsoft Windows Azure Mobile Services, AWS Mobile Services). The main available services for development of mobile cloud-based applications are presented in Table 1.

The examples of mobile back end as a service described in Table 1 allow for developing and deploying web-based, native, or hybrid mobile applications and running them on multiple devices. Learners can have different mobile platforms to use cloud-based m-learning applications. Furthermore, if the Internet is temporally unavailable, learners still can continue working on their mobile devices locally and the changes made will be synchronized in the cloud when Internet will be available.

The Push Notification service allows to easily pushing data to the right users at the right time on the mobile devices. Notifications can be sent to a single device, or a group of devices based on their subscriptions.

Depending on the learning scenario this service can be used to support individual learner or group of learners while performing a certain learning activity. In addition to the Push Notification service, IBM's BlueMix mobile cloud service offers a number of services that can be used to develop m-learning applications.

For example, the Language Identification service allows detecting the language in which input text is written while the Machine Translation service enables to

Table 1 Examples of mobile cloud-based services

Cloud provider	Platform SDK's	Database	Analytics	Cloud functions/services
Microsoft windows Azure mobile services	Windows Phone 8, Android, iOS, HTML5	SQL, MongoDB	Mobile analytics with captain	Push notifications
AWS mobile services	Android, iOS	Amazon DynamoDB	Custom analysis, Amazon mobile analytics reports	Push notifications, data streaming, AWS Lambda
Oracle cloud mobile services	Android iOS	Oracle database	Oracle analytics	Push notifications
IBM bluemix mobile cloud services	Android, iOS, hybrid, node.js	DB2, cloudant NoSQL DB, SQL	IBM embeddable reporting, SPSS analytics	Push notifications, language identification, machine translation, personality insights, speech to text, text to speech, visual recognition, presence insights, and more

translate text from one language to another. This latest service can be useful for supporting a multi-language communication on forums where a student can write a text message on his/her native-speaking language and its language will be automatically identified and translated to the student's receiver native-speaking language.

Service providers can add and expand their service offerings. Multiple services from different providers can be integrated easily through the cloud to meet today's complex user demands and increase users' engagements with m-learning applications.

3.3 Advantages of Mobile Cloud-Computing Services

In this subsection, we describe how mobile cloud-computing services can be used to support learning activities, to overcome obstacles related to m-learning and to enhance learners' engagement with m-learning applications:

Context-aware learning. Supports context-aware learning activities for learners. For example, providing learning resources for learners by recommending the appropriate content to users based on an intelligent analysis of the learners' behaviors and their learning outcomes [27]. The CloudAware framework offers context adaptation features through *Jadex* middleware with a combination of agent-, service-, and component-oriented engineering perspectives [28].

Another interested feature for mobile application development is the creation of a function that is executed in response to an event (e.g., notifications, messages, image uploads) and it has been introduced in AWS Lambda Adds. Those functions are written in NodeJs framework that invoked in synchronous manner and receive the context information of the application data (e.g., name, build, version, and package), device data (e.g., manufacturer, model, platform), and user data (e.g., a client id) as part of the request. This feature allows to easily create rich, adaptable and personalized responses to in-app activities.

Security. The mobile application security service provides or blocks any devices and/or users by using additional user authentication [29]. Moreover, it provides possibility to configure the access, sharing settings and protect personal data. Most CC providers offer flexible and reliable backup and recovery solutions [30].

Accessibility. The m-learning systems typically include different kinds of multimedia resources helping learners to be more engaged and interested in collaboration [31, 32]. The efforts carried out by [32] provide the Learning Cloud framework, where users can work on different operating systems for mobile devices, and in that way students and teachers can access the cloud-based platform simultaneously from any location, at any time.

Another study carried out by [31] has used a cloud-based learning platform to support distance learning and to provide an increased quality of e-learning. In m-learning scenarios like field trips, where students are taking photos by using their mobile cameras to collect some information and data about the learning environment, it is necessary to have some storage capacity and search/retrieve mechanism.

Here, for example, the photos can be stored and processed directly into the cloud instead of mobile device [33]. There are also advantages for teachers having cloud-based applications to manage everything from documents to students' attendance and grades [34]. For example, with the TeacherKit application teachers can organize classes and manage students' activities easily. With the SchoolTube application teachers can upload educational videos for students.

Another example is the Edmondo application where teachers, students, and parents are connected to collaborate on assignments, discover new resources, and more. However, there are still no cloud-based services that provide possibilities for teachers to design their own m-learning applications and to distribute them to students on multiply mobile platforms.

Cloud services for learning scenarios can be used for both in formal and informal learning settings, remote, and colocated situations and in synchronous or asynchronous scenarios [35]. The combination of different cloud services, sensors information, storage capacity, and cloud computation power provides new opportunities to develop flexible and highly personalized cross-platform m-learning applications.

3.4 Limitation

In this section, we presented an overview with regard to the possibilities and potential that different CC environments and methods can have within the field of TEL.

We are aware that a lot of different applications and services are listed within this section. To describe each one of them in more detailed fashion would be out of scope of this chapter, therefore, we leave it to the reader to investigate the mentioned applications/services/methods and leave the description at the current abstraction level.

4 Contextualized Cloud-Based Services and Web-Based Authoring Framework

In order to address our research questions, we present in this section two cloud-based solutions that are strongly connected. We present first a flexible *Web-based Framework* to enable teachers to compose and deploy their own mobile applications to perform m-learning activities. Additionally, we present a *Contextualization Service* that has been developed in order to improve the content delivery of LOs. This service allows adapting the representational format of the learning content and the mobile UI to the current context of the user.

4.1 Web-Based Framework

There are learning activities that are connected to tasks that include data collection, analysis and visualization; our framework offers to take advantage of the internal sensors available at modern mobile devices. The fact that the framework is realized in a cloud-based environment, tackles the previously mentioned challenges in terms of scalability and reliability. The presented web-based framework, *mLearn4web*, comes with an authoring tool, where teachers can design mobile applications for instance for field trips scenarios. Additional to the authoring tool, mLearn4web consists of two more components that are all purely based on web-technologies: a mobile component to perform learning activities and a visualization component that offers analyzing methods of the data that has been generated by the mobile component.

A major challenge in the field of TEL is the fact that teachers often do not have the technical skills that are required to create applications that can fit their pedagogical needs. Often, they need to consult with researchers and/or developers to create or adjust applications to their specific requirements, which are not only inconvenient for the teachers but also generate additional work for researchers/developers.

Therefore, the mentioned authoring tool allows designing mobile applications by using simple and well-known interaction methods like drag and drop. In our case, the mobile applications consist of a number of screens and the authoring tool can modify the content and functionalities within a screen. The authoring tool is divided into three areas: a screen area; a content area; and an element area. In the screen area, users can add/delete screens and change their appearance order by dragging a screen to a new position. Users can add functionalities and content to screens of a mobile application by dragging predefined elements from a list to the content area. This includes the access of certain internal sensors of mobile devices like camera, microphone, or Global Positioning System (GPS).

Furthermore, it is possible to add the following elements: an instruction that allows providing text information; a text area that allows users to enter text on a mobile device; a multiple choice element where it is possible to pose a multiple choice questions and the user can pick an answer at the mobile device; a numerical input field; and a date input field that allows to enter a date in the proper format. Figure 1 illustrates the functionalities of the authoring tool.

The authoring tool allows even nontechnical skilled users to easily generate mobile applications that have the functionalities fitting their needs. After the design process is finished a mobile application is automatically deployed and available as a web application. The fact that all components are based on web-technologies allows the easy deployment in CC environments like OpenShift. Since no developers/ researchers are involved in the process, the potential of having a huge amount of mobile applications is given. Therefore, the mobile application is deployed in a CC environment. Figure 2 shows examples of how the mobile application looks like.

We have conducted several studies [36–38], and the outcomes of our efforts indicate that teachers without technical knowledge could generate and deploy mobile application fitting their specific needs.

Of particular interest is the fact that even though some teachers had troubles in the beginning to deal with the functionalities of the framework, managed at the end to design and deploy mobile applications and repeated the process without having

Fig. 1 Screenshot of the authoring tool

Fig. 2 Examples of the mobile application

much troubles [36]. This shows that the system is not only easy to use but also has a high learnability factor.

The third component of mLearn4web (Fig. 3), the visualization component, processes data that is generated during the usage of the mobile application. This allows teachers and students to reflect upon learning activities performed with the mobile application.

Datasets that are generated by the mobile application can be brought into context to each other. For instance, if GPS data and a picture are collected at the same screen, it is likely that they are contextually connected. Therefore, a map is offered, where the marker shows the picture that is also available. However, there is more potential for presenting contextualized information.

It is, for example, possible to use existing web services to gather more information and to aggregate to the data generated by the mobile application.

Fig. 3 The visualization component of mLearn4web

For instance, if GPS data is available, it is possible to add information about the surrounding environment or weather to the visualization.

4.2 Contextualized Service

The *Contextualized Service* is a service that provides personalization for mobile applications to the current context of the user. Particularly, it supports personalized interactions with a mobile device in order to meet the user's needs and goals. The main goal of personalization is to improve the user experience by taking advantages of contextual information in order to provide adequate services.

Examples of contextualized services are location based services [39] that provide useful and relevant information to the current users' location; contextualized knowledge services [40] for supporting learners in a personalized and adaptive way by using the context information; contextualized learning services [41] for providing a personalized feedback in learning support. Based on our previous research work [42], the contextualization of learners could vary from one learning scenario to another.

Unlike in a traditional learning environment, in an m-learning environment, interactions with the learning applications are performed across a variety of contexts. Thus, it is important to identify the learners' needs, goals, and expected outcomes to provide them with contextualized services. For example, in a location-related learning scenario the relevant content should be provided to the learner at the right place and time (e.g., museums, field trips). Below we describe three features that personalized m-learning applications should have while providing a contextualized service.

Contextual Content Representation. This feature allows adapting the format of LOs to the current learner's situation. We describe the learner's situation by contextual data gathered from mobile sensors, additional external Web Service API's, social networks and store it in a database in the cloud. Utilizing a cloud service with large storage capacity for collecting, storing, and processing the contextual data provides the possibility to use not only real-time contextual data but also historical users' contextual data that can improve the prediction of the format of LO that is best suited to the current context of the learner.

For predicting the relevant content of the LO the proposed flexible *rich context model* (RCM) [42] has been used. The main difference of having RCM on the cloud is that allows to store and process a vast amount of contextual data, to support a large number of mobile users and to use additional cloud-based services to enrich the contextual data. This feature can be used in different m-learning scenarios to provide convenient support to the learning processes anywhere and at anytime.

Contextualized User Interface. This feature allows adapting UI elements of a web-based mobile application to the current context of the user (e.g., light/dark colors in the themes, predictable input form elements, and adaptable UI elements).

This allows making the interaction with a mobile device adaptable to the current context of the user.

For example, provide convenient volume of the mobile device (e.g., make it lower/upper) by taking into account the noise level of the device environment and position of the device, user's activity, place, and time. Another good example is an application called Star Walk[1] application, which uses mobile camera, compass, location, and augmented reality that allows learning and exploring the information about the universe by holding the phone at the night sky. Another example can be a contextualized keyboard for learning chemistry through mobile application (e.g., ChemCalc[2]) or for having personalized keyboard application called SwiftKey[3] that delivers smart predictions and fast typing. Another example is a personalized application launcher that organizes the application bar with the most used mobile applications in a certain time, place, and day (e.g., the user is at a school then the application bar shows only thus applications that was often used by user at the school).

Contextual Notifications. This feature allows users to make their mobile devices more personalized according to their current needs and interests. For example, a student is interested on buying a particular thing, and then the application sends him/her the discounts/offers related to this object, depending on the users' leaving place for taking into account the delivery costs or using the local shops. Additionally, reminders of duties, course schedule, etc., can be sent as notifications to the mobile device in a suitable format (e.g., the voice notification, the image notification, and the text notification). In the case of m-learning scenarios, this feature can be used for guiding students in learning environment such as field trips or visits to thematic parks or museums. The described three features above can be implemented as a set of micro-services and integrated in the micro-service architecture as shown in Fig. 4.

It is shown in Fig. 4 that each feature is implemented as a micro-service and it has its own database in order to use it independently from the scenario point of view. The micro-services architecture is an approach for developing an application or service as a set of small independent services [43]. The main advantages of using a micro-service approach are scalability (e.g., scaling a certain feature instead of whole application), maintainability (e.g., easy to maintain since each micro-service implements a single feature/functionality), flexibility in distribution of the resources for each micro-service (e.g., the single function as *analyze contextual data* will need more computational recourses then the *collect contextual data* function) and extendibility in adding features/functions by increasing/adding/implementing micro-services.

Then, different features provided by the contextualized service can be used independently from the scenario point of view. For example, one learning scenario

---

[1]https://itunes.apple.com/en/app/star-walk-5-stars-astronomy/id295430577?mt=8.

[2]https://itunes.apple.com/us/app/chemcalc/id499955745?mt=8.

[3]https://swiftkey.com/en/.

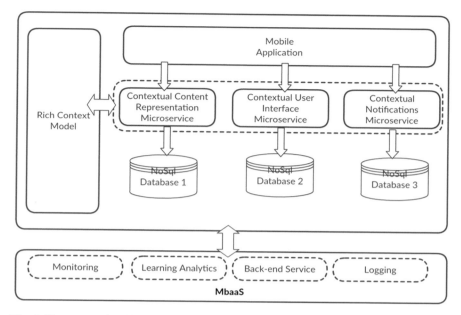

Fig. 4 The proposed architecture overview for contextualized service

can use only the *contextual content representation* feature while another one only *contextual notifications* depending of the nature of the activity. Furthermore, the service is flexibly in a way that allows adding new features as an independent micro-service and uses it in combination with others. The proposed service is deployed in a cloud environment for supporting contextualization of mobile users in real time. In our case, the *RCM* has been utilized [42] to support these features.

4.3 Rich Context Model

The *RCM* we are using is a model that handles the rich context of a mobile user. Here, the rich context is a data set received from different mobile sensors available on the device. Moreover, this data can be enhanced by using external Web Services (e.g., Google Places API) to retrieve more detailed information about the context of the mobile user (e.g., weather condition, nearby services) [42].

An example of different context dimensions and mobile sensors that can be used in different m-learning scenarios is given in Table 2. There is no use for sensors if the data will not be processed and analyzed. Context analysis techniques allow processing these data and providing a meaningful interpretation. The available contextual data is evaluated by the use of a multidimensional approach that provides richer representation of the user's context [44]. Each dimension describes a

Table 2 Example of context dimensions and mobile sensors in m-learning scenario

Context dimension	Mobile sensors	Contextual information	Web services
Environment context	GPS	The type of place where the learner is located; the nearest places around the learner's location; the weather conditions when the learning environment is outdoor	Google places API, free weather API
	Accelerometer	The learner's status (sitting in the train, bus, etc.)	
	Digital compass	The learner's direction	
	Camera	Fast and easy access to the learning materials by scanning barcodes, qrcodes; the type of place by taking picture of place where learner is located	QRCode web service
Device context		Screen size, battery charge, internet connectivity	
Personal context		Interests, language, country	Facebook API, Twitter API

property of an entity or the entity itself, where the entity can be an object, a person, or a situation.

This enables the RCM to consider various data types of the contextual information and to use different approaches to evaluate and analyze the data. Another example of what can be achieved with the RCM is the identification of a context similar to the user's current context in order to provide relevant recommendations. In addition, a suitable format for the representation of the learning materials can be recommended with the RCM by taking the environmental context information into account. Here, the LOs in different formats are described by contextual information provided in Table 2 and represented in *multidimensional vector space model* (MVSM) as shown on Fig. 5.

Then, in order to define the best-suited format of a LO, the distance between the vector of the current context of the user and the vectors at the MVSM is calculated by using the combination of different metrics similarity (cosine distance, Euclidean distance, Jaccard distance, etc.). The vector that has minimal distance to the vector of current context of the user defines the best-suited format of LO.

The context information has various data types, and therefore, different metrics similarity or algorithms should be considered to process this data. As an example, we used the combination of different metrics and algorithms that are shown in Table 3.

Having a generic approach for context modeling is very important, because different learning scenarios may have dissimilar properties describing an entity itself and therefore different dimensions in a context model.

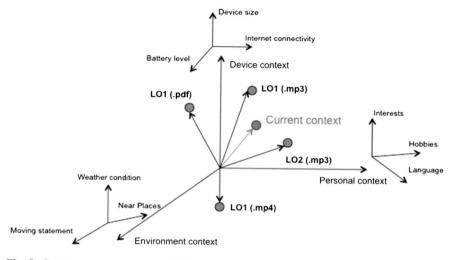

Fig. 5 Context representation in MVSM

Table 3 Examples of usage different metrics and algorithms for processing contextual information

Contextual information	Data type	Metric similarity/algorithm
Hobbies	Array of Boolean (1—has hobby, 0—does not have hobby)	Jaccard distance
Age	Integer	Euclidean distance
Movement statement, outside/inside the building	Boolean	Jaccard distance
Additional information	Long text/string	Latent semantic analysis

A technical infrastructure to support this is outlined in Fig. 6. The major task performed by this infrastructure is to provide a certain abstraction for the context dimensions in order to allow an automatic and flexible configuration in the RCM.

In this architecture, the *Context Modeling* component is responsible for defining the context dimensions that should be used for a certain scenario. The web-based application proposed to configure these context dimensions and can be used by an expert. We define *an expert* as a person (e.g., teacher) that has a wide body of knowledge in relation to subject matter that is at the core of the learning scenario in which the contextualized service will be used.

The developed web application allows visually adding/removing context dimensions and subdimensions. After adding all the required dimensions into the scenario, the expert can export the context modeling results into a JSON file. This JSON file is an input for our contextualized approach and it is required for the

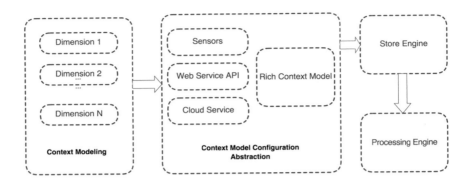

Fig. 6 Architecture overview for RCM configuration

usage of the contextualized Service. Then the *Context Model Configuration* block has an abstraction for the context dimensions and it is responsible for the configuration of RCM. After the RCM is configured it is used for collecting, storing and processing contextual data. We think that an expert in a certain learning scenario should define the context dimensions as he/she has best knowledge about it. This makes our contextualized service flexibly in terms of supporting different context models for learning scenarios.

The model supports different recommendation types, e.g., relevant LOs and convenient representational format of LOs, and can therefore be extended by others types of recommendations. Additionally, it supports different levels of granularity of the context. This is achieved by the flexibility of the RCM to include/exclude the context dimensions. It supports using priorities for different contextual information. The latest is achieved by using additional weights for each context sub-dimension according to the user's preferences. This in theory allows the RCM to provide better and more convenient services for mobile users. The main difference between our RCM and similar approaches is the flexibility that it allows for describing a complex contextual situation. Our approach has the potential to improve the usability of mobile devices in order to enrich the current context of the users.

For some learning scenarios it might be the case when an expert does not know which context dimensions should be considered in order to achieve good recommendation results. We suggest two possible solutions (a) to perform a pre-study in order to identify and evaluate the necessary context dimensions used in our context model. Since the proposed approach uses the comparison of the current context of the user with objects that were consumed by others in a similar context situation, the pre-executed study should be performed in order to have the database of different context situations; (b) to collect contextual data for different context's situations and apply machine learning algorithms (e.g., vector support machine) to define which contextual data should be used to classify different contexts' situations for a certain learning scenario.

5   Scenario Examples

This section presents two scenario examples: (a) using the *mLearn4web* framework in an m-learning scenario and (b) using the Contextualized Service with the *LnuGuide* mobile application.

The *mLearn4web* is a web-based framework that allows users without programming skills to design and deploy m-learning activities [36]. The *LnuGuide* is mobile application that supports exchange students while they are explore the university facility services on the campus. The purpose of these designed scenarios is to outline the advantages and benefits for learners and teachers while using the *mLearn4web* framework with the *Contextualized Service* in a cloud environment.

5.1   mLearn4web

The *mLearn4web* is a web-based framework designed in [37] for creation and deploying m-learning activities. It consists of three main components: an authoring tool that was described in Sect. 4.1, a mobile application and a visualization tool. These components can be used to support the three phases that are described in [45] which together provide one of the most prominent examples of a m-learning activity: a field trip. The authoring tool offers support for the "pre-trip phase," where the preparation of the field trip takes place.

The mobile application offers support to the actual field trip. The visualization tool supports the "post-trip phase" in which the debriefing and analysis of the field trip takes place. In this field-trip scenario, these are the units of interest for the RCM. For instance, in the "pre-trip phase" the teacher defines which kind of data should be collected in a certain learning activity. Then, these collected data (e.g., images, location, textual comments, numerical values) could be considered as a part of the contextual information and stored at the RCM.

Here, the RCM will represent the collected data by contextual information as described in Table 2. Then, the result of the contextualization will be the delivery of interested and relevant information about the learning environment to the current context of the learner. For instance, when the learner will reach a certain place, the application will show the most relevant objects and data gathered by other learners in the similar context situation. This might allow for learners to find out more information about the learning environment or do not miss important and useful pieces of information that are related to a location/place. This information can be delivered with the help of contextual notification services provided by the contextualized service.

In order to tackle the issues related to scalability, we have deployed the mLearn4web in a cloud environment. Having the *mLearn4web* application in a cloud allows automatically deploying the resulting designed m-learning activities into the cloud. This may turn out to a powerful and flexible framework for the provisioning of (potentially ad hoc) learning scenarios.

5.2 LnuGuide

M-learning activities can be designed for guiding mobile learners [46] to gain information about current learning environment and how to work in it. For instance, students can learn how to use the different services at the university library (e.g., registration at the library, using the library card, etc.) if he/she is on site.

Another example might be that students can be guided to learn how to print and scan articles by using the university printing system. The *LnuGuide* m-learning scenario was designed and developed for exchange students to get familiar with campus and prominent institutions and services on it [42]. The LnuGuide activity contains three stations (e.g., University Library, Administration Building and a café on campus) where students can get useful information to facilitate his/her "student life" (e.g., obtain the library or student card, to be able to scan and print at Library, etc.). Each station provides a number of tasks, where for instance the app will provide information on how to scan documents at the library including instructions that the user should easily be able to perform.

In this particular scenario, the contextualization of learners is supported by recommending a convenient format of learning material that is suitable to the learner's current context. Each learning material has been represented by different formats (MP3, PDF, PPTX, MP4) and described by contextual information in the RCM. Here, the following three main context dimensions were used: *personal*, *device* and *environment* context.

Additional cloud-based services as *Speech to Text* and *Text to Speech* (Table 1) Services can be used to provide convenient representation format of LO to mobile device instead of having and storing the LOs in an audio format (e.g., MP3). The LnuGuide application requires having reliable Internet connection during the performed learning activity. Therefore, the cloud-based offline support service can be used to overcome this issue and increase the student's concentration level on performing a certain task without interruption (e.g., Internet connection not available).

Additionally, the exchange students come form different countries and therefor specks different languages. Then by using a Language Identification Cloud Service, the LnuGuide application can be adapted to the students' native specking language. Moreover, the analyses of log files can be replaced by the use of rich analytics cloud services for monitoring the students' performance. These described features provide possibilities to improve the usability and personalization of the LnuGuide application.

5.3 *Responses for the Research Questions*

In order to address the research questions formulated in the beginning of the chapter two scenario examples of using (a) the mLearn4web framework and (b) the Contextualization Service in a cloud were proposed. These examples show that the cloud-based solutions have the potential to improve the contextualization approach by adding new available cloud services to provide more highly personalized applications.

Furthermore, storing more contextual data to improve recommendations, by monitoring user's interactions with application to increase their engagement with the app and the understanding of users' current needs can also help to increase the level of personalization. Additionally, it allows using different algorithms (e.g., machine learning algorithms) for processing contextual data and providing recommendation to the user in real time.

The described examples also show that the most useful cloud-based services, such as the Learning Analytics Service, Logging Service, and Monitoring Service provide new opportunities for designing novel services. All of them provide the possibility to gather more information about the current context of the learner in order to provide him/her better recommendations.

6 Conclusions

This chapter described and discussed different cloud-based services and solutions used for educational purposes. These can provide contextual support for mobile learners. Additionally, we presented the most popular cloud-based applications and services used by teachers and learners. Overall, this chapter presented (a) a flexible web-based framework, mLearn4web, to enable teachers to compose and deploy their own mobile applications to perform m-learning activities and (b) a contextualized service to improve the content delivery of LOs on mobile devices. This service allows for adapting: (1) the learning content and (2) the mobile UI to the current context of the user. For the realization of the contextualization approach, a RCM deployed in a cloud-based environment has been utilized. The combination of the two cloud-based approaches described in this chapter provides a powerful flexible framework that can enhance the learning experience for both, teachers and students.

Our future efforts toward the refinement and improvement of contextualized services include the development of additional m-learning applications with a focus on personalization and the use of data analytics to increase students' motivation, performance, and engagement. Current CC technologies and services offer new possibilities for supporting the contextualization of users and for providing personalized and adaptable services and applications that can enhance m-learning activities.

References

1. Ruth, S., Mason, G.: Is e-learning really working? The trillion-dollar question (2010)
2. Berge, Z.L., Muilenburg, L.: Handbook of Mobile Learning. Routledge (2013)
3. O'Malley, C., Vavoula, G., Glew, J., Taylor, J., Sharples, M., Lefrere, P., Lonsdale, P., Naismith, L., Waycott, J.: Guidelines for Learning/Teaching/Tutoring in a Mobile Environment (2005)
4. Farmer, J., Knapp, D., Benton, G.M.: An elementary school environmental education field trip: long-term effects on ecological and environmental knowledge and attitude development. J. Environ. Educ. **38**, 33–42 (2007)
5. Shih, J.L., Chu, H.C., Hwang, G.J., Kinshuk, : An investigation of attitudes of students and teachers about participating in a context-aware ubiquitous learning activity. Br. J. Educ. Technol. **42**, 373–394 (2011)
6. Bosch, J.: From software product lines to software ecosystems. In: Proceedings of 13th International Software Product Line Conference, pp. 111–119 (2009)
7. Jansen, M., Bollen, L., Baloian, N., Hoppe, H.U.: Using cloud services to develop learning scenarios from a software engineering perspective an alternative perspective on cloud services for learning scenarios different categories for cloud services for learning scenarios. J. Univers. Comput. Sci. **19**, 2037–2053 (2013)
8. Martin, J.P., S., C.K., M.J., H., S., A.B., Cherian, S., Sastri, Y.: Learning environment as a service (LEaaS): cloud. In: 2014 Fourth International Conference Advanced Computing Communicatyion, pp. 218–222 (2014)
9. Lefever, R., Currant, B.: How can technology be used to improve the learner experience at points of transition. High. Educ., Acad (2010)
10. Velev, D.G.: Challenges and opportunities of cloud-based mobile learning. Int. J. Inf. Educ. Technol. **4**, 49–53 (2014)
11. González-Martínez, J.A., Bote-Lorenzo, M.L., Gómez-Sánchez, E., Cano-Parra, R.: Cloud computing and education: a state-of-the-art survey. Comput. Educ. **80**, 132–151 (2015)
12. Verma, K., Dubey, S., Rizvi, M.A.: Mobile cloud a new vehicle for learning : m-learning its issues. Int. J. Sci. Appl. Inf. Tech. **1**, 93–97 (2012). ISSN No . 2278-3083
13. Kitanov, S., Davcev, D.: Mobile cloud computing environment as a support for mobile learning. In: Third International Conference Cloud Computing GRIDs, Virtualization Mob, pp. 99–105 (2012)
14. Gupta, N., Agarwal, A.: Context Aware Mobile Cloud Computing : Review. 1–5
15. Wang, S., Dey, S.: Adaptive mobile cloud computing to enable rich mobile multimedia applications. IEEE Trans. Multimed. **15**, 870–883 (2013)
16. Rao, N.M.: Cloud computing through mobile-learning. Int. J. **1**, 42–47 (2010)
17. Thomas, P.Y.: Cloud computing: a potential paradigm for practising the scholarship of teaching and learning. Electron. Libr. **29**, 214–224 (2011)
18. Galih, S.: Mobile cloud based learning material repository using android and google drive application. In: Second International Conference Digit. …. pp. 80–83 (2013)
19. Sun, G., Member, S., Shen, J., Member, S.: Facilitating social collaboration in mobile cloud-based learning : a teamwork as a service (TaaS) approach. IEEE Trans. Learn. Technol. **7**, 207–220 (2014)
20. Lin, Y.T., Wen, M.L., Jou, M., Wu, D.W.: A cloud-based learning environment for developing student reflection abilities. Comput. Hum. Behav. **32**, 244–252 (2014)
21. Siemens, G., d Baker, R.S.J.: Learning analytics and educational data mining: towards communication and collaboration. In: Proceedings of the 2nd International Conference on Learning Analytics and Knowledge, pp. 252–254 (2012)
22. Madjarov, I., Normandie-niemen, A.E., Architecture, A.C.C.: Cloud-Based Framework for Mobile Learning Content Adaptation. 381–386 (2014)

23. Zurita, G., Baloian, N., Frez, J.: Using the cloud to develop applications supporting geo-collaborative situated learning. Futur. Gener. Comput. Syst. **34**, 124–137 (2014)
24. Ferguson, R.: Learning analytics: drivers, developments and challenges. Int. J. Technol. Enhanc. Learn. **4**, 304–317 (2012)
25. Tabot, A.: Mobile Learning with Google App Engine (2014)
26. Zhao, W., Sun, Y., Dai, L.: Improving computer basis teaching through mobile communication and cloud computing technology. In: Proceedings of ICACTE 2010–2010 3rd International Conference Advanced Computing Theory Engineering, vol. **1**, pp. 452–454 (2010)
27. Wang, M., Ng, J.W.P.: Intelligent mobile cloud education: smart anytime-anywhere learning for the next generation campus environment. In: Proceedings of 8th International Conference Intelligent Environment IE 2012, pp. 149–156 (2012)
28. Orsini, G., Bade, D., Lamersdorf, W.: CloudAware : Towards context-adaptive mobile cloud computing. In: IFIP/IEEE IM 2015 Workshop: 7th International Workshop on Management of the Future Internet (ManFI), pp. 1–6 (2015)
29. Dinh, H.T., Lee, C., Niyato, D., Wang, P.: A survey of mobile cloud computing: architecture, applications, and approaches. Wirel. Commun. Mob. Comput. **13**, 1587–1611 (2013)
30. Alzahrani, A., Alalwan, N., Sarrab, M.: Mobile cloud computing. In: Proceedings of 7th Euro American Conference Telematics. Information System—EATIS '14, pp. 1–4 (2014)
31. Karadimce, A., Davcev, D.: Experiments in Collaborative Cloud-based Distance Learning, pp. 46–50 (2013)
32. Chang, C.S., Chen, T.S., Hsu, H.L.: The implications of learning cloud for education: from the perspectives of learners. In: Proceedings of 2012 17th IEEE International Conference Wireless, Mobile. Ubiquitous Technology Education WMUTE 2012, pp. 157–161 (2012)
33. Cai-dong, G., Kan, L., Jian-ping, W., Ying-li, F., Jing-xiang, L., Chang-shui, X., Mao-xin, S., Zhao-bin, L.: The investigation of cloud-computing-based image mining mechanism in mobile communication WEB on android. In: 2010 Ninth International Conference Grid Cloud Computing, pp. 408–411 (2010)
34. Walling, D.R.: Do you moodle? In: Designing Learning for Tablet Classrooms, pp. 119–125. Springer, Berlin (2014)
35. Jansen, M., Bollen, L., Baloian, N., Hoppe, H.U.: Cloud services for learning scenarios: widening the perspective. CEUR Workshop Proceedings **945**, 33–37 (2012)
36. Zbick, J., Jansen, M., Milrad, M.: Towards a web-based framework to support end-user programming of mobile learning activities. In: IEEE 14th International Conference on Advanced Learning Technologies (ICALT), pp. 204–208 (2014)
37. Zbick, J., Nake, I., Jansen, M., Milrad, M.: mLearn4web. In: Proceedings of 13th International Conference Mobile. Ubiquitous Multimedia—MUM '14, pp. 252–255 (2014)
38. Zbick, J., Nake, I., Milrad, M., Jansen, M.: A web-based framework to design and deploy mobile learning activities: evaluating its usability, learnability and acceptance. In: 15th International Conference on Advanced Learning Technologies (ICALT 2015), pp. 88–92 (2015)
39. Liu, Y., Wilde, E.: Personalized location-based services. In: Proceedings of the 2011 iConference, pp. 496–502 (2011)
40. Rath, A.S., Devaurs, D., Lindstaedt, S.N.: Contextualized knowledge services for personalized learner support. In: Proceedings of Demonstrations EC-TEL'09 (2009)
41. Specht, M.: Towards contextualized learning services. In: Learning network services for professional development, pp. 241–253. Springer, Berlin (2009)
42. Sotsenko, A., Jansen, M., Milrad, M.: Implementing and validating a mobile learning scenario using contextualized learning objects. In: The 22nd International Conference on Computers in Education (ICCE), November 30, 2014 to December 4, 2014, Nara, Japan, pp. 522–527 (2014)

43. Namiot, D., Sneps-Sneppe, M.: On micro-services architecture. Int. J. Open Inf. Technol. **2**, 24–27 (2014)
44. Sotsenko, A., Jansen, M., Milrad, M.: About the contextualization of learning objects in mobile learning settings. In: 12th World Conference on Mobile and Contextual Learning (mLearn 2013), pp. 67–70 (2013)
45. Krepel, W.J., DuVall, C.R.: A Study of School Board Policies and Administrative Procedures for Dealing with Field Trips in School Districts in Cities with Populations over 100,000 in the United States (1972)
46. Sharples, M., Others: Big issues in mobile learning: report of a workshop by the kaleidoscope network of excellence mobile learning initiative. University of Nottingham, pp. 1–37 (2007)

Toward an Adaptive and Adaptable Architecture to Support Ubiquitous Learning Activities

Janosch Zbick, Bahtijar Vogel, Daniel Spikol, Marc Jansen
and Marcelo Milrad

Abstract The continuous evolution of learning technologies combined with the changes within ubiquitous learning environments in which they operate result in dynamic and complex requirements that are challenging to meet. The fact that these systems evolve over time makes it difficult to adapt to the constant changing requirements. Existing approaches in the field of Technology Enhanced Learning (TEL) are generally not addressing those issues and they fail to adapt to those dynamic situations. In this chapter, we investigate the notion of an adaptive and adaptable architecture as a possible solution to address these challenges. We conduct a literature survey upon the state of the art and state of practice in this area. The outcomes of those efforts result in an initial model of a Domain-specific architecture to tackle the issues of adaptability and adaptiveness. To illustrate these ideas, we provide a number of scenarios where this architecture can be applied or is already applied. Our proposed approach serves as a foundation for addressing future ubiquitous learning applications where new technologies constantly emerge and new requirements evolve.

Keywords Domain-specific architecture · Web-based · Adaptive and adaptable architecture · Technology enhanced learning · Ubiquitous learning

J. Zbick (✉) · M. Jansen · M. Milrad
Department of Media Technology, Linnaeus University, D-Building, Vejdes plats 6–7,
351 95 Växjö, Sweden
e-mail: janosch.zbick@lnu.se

M. Jansen
e-mail: marcbjoern.jansen@lnu.se

M. Milrad
e-mail: marcelo.milrad@lnu.se

B. Vogel · D. Spikol
Department of Media Technology, Malmö University, Nordenskiöldsgatan 1,
205 06 Malmö, Sweden
e-mail: bahtijar.vogel@mah.se

D. Spikol
e-mail: daniel.spikol@mah.se

© Springer International Publishing Switzerland 2016
A. Peña-Ayala (ed.), *Mobile, Ubiquitous, and Pervasive Learning*,
Advances in Intelligent Systems and Computing 406,
DOI 10.1007/978-3-319-26518-6_8

Abbreviations

CCOA	Cloud computing open architecture
DSA	Domain-specific architecture
DSSA	Domain-specific software architecture
GEM	Geometry mobile, a project by the CeLeKT research group at LNU
ICT	Information and communication technologies
LCD	Learner-centered design
LETS	Learning ecology through science with global outcomes, a project by
GO	the CeLeKT research group at LNU
NASA	National Aeronautics and Space Administration
PELARS	Practice-based experiential learning analytics research and support, an EU project
SA	Software architecture
TEL	Technology enhanced learning

1 Introduction and Motivation

The rapid and constant evolution of web and mobile technologies brings new opportunities to researchers and developers in the process of creating innovative mobile applications. Internal sensors that allow generating contextualized data as well as ubiquitous Internet access have become standard features of modern mobile devices. These extended features provide new possibilities for augmenting learning activities. The mentioned potentials can incorporate also different physical and environmental sensor data [1] and different web, mobile, and sensor-based technologies can provide new perspectives on how learning activities can be embedded in different settings and across contexts [2, 3]. One innovative aspect of these new learning landscapes is the combination of learning activities to be conducted across different educational contexts, such as schools, nature and science centers/museums, parks, and field trips [4]. In these ubiquitous learning environments users make use of a wide range of devices, services and applications, and the notions of an adaptive and adaptable architecture becomes central for our efforts.

Coping with heterogeneity of the technologies and activities in this domain, the adaptive and adaptable architecture should facilitate the design and development of flexible solutions. This type of architecture should be extensible as well as evolvable so that it can be easily adapted to the domain-specific characteristics of each application domain and scenarios.

In the remaining of this chapter, we explore the mentioned notions in order to tackle the challenges posed by the different educational activities, the evolving needs of teachers and learners as well as the rapid evolution in technologies.

2 Background and Research Questions

The continuous evolution of learning technologies combined with the changes within ubiquitous learning environments in which they operate result in dynamic, complex requirements that are challenging to satisfy. As a result, the evolution of web and mobile software, together with the dynamic requirements generated, and the fact that these systems evolve over time, make it difficult to adapt to the constant changing requirements [5, 6]. Furthermore, the process of design needs to be extended to accommodate the shift from designing objects to more socio-material experiences [7, 8].

Proprietary solutions have been deployed extensively throughout multiple platforms, including desktop, web, and mobile systems, which in many cases are closed, which means they can only be adapted to new requirements or technologies by the owners of those solutions. Therefore, it is difficult for those systems to evolve the design of the experience for specific learning activities.

Web technologies offer a promising approach to tackle many of the mentioned challenges. However, while working with web technologies, in particular regarding mobile development, processes are becoming fragmented; there are multiple browsers, which in different ways comply with web standards; diverse mobile platforms, where the operating systems and programming languages are different; and physical devices that have varying characteristics [9–11].

Ubiquitous learning faces the challenges of rapid response to adapt to the need of users and quick technological changes as the learning ecosystem evolves. This requires the ability for a rapid modification and extension of the solution. Therefore, solutions in this area have to have the ability to grow and evolve over time. An architecture is desirable that represents a fundamental organization of a system in terms of the key components, their relationships to each other, and to the environments, which reflects the rationale behind the system's structure, functionality, and interactions [12].

We must also emphasize the growing demands for providing rich user experiences with web and mobile applications. As such, rich user experiences become more important due to the significance of the user context [13, 14]. Therefore, in these environments, architectural aspects play a crucial role [15]. Additionally, the aspects of designing for ubiquitous learning need to work in harmony between the adaptable architecture and the educational needs.

Due to the evolving requirements in TEL and technological settings it becomes a central challenge to provide an adaptive and adaptable architecture. As sophisticated technologies constantly emerge and in order to support adaptation to new requirements and the use of context over time [6] concerns defining what would be an acceptable adaptation behavior as well. To the best of our knowledge there is not much work conducted to investigate an architectural approach to address these challenges.

In summary, the interplay between closed systems, that are difficult to adapt to new requirements, fragmentation of devices and operating systems in ubiquitous learning environments, and changing requirements, and therefore, the need for systems to have the ability to adapt and grow over time provide fundamental architectural challenges that extend into user experience design. This architectural approach becomes relevant particularly when dealing with dynamic and ubiquitous learning environments. Therefore, our current efforts are directed toward exploring the following research questions:

- (RQ1) How to provide an adaptive and adaptable architecture in order to fulfill constantly evolving requirements in dynamic and ubiquitous learning environments?
- (RQ2) What are the general requirements for an adaptive and adaptable architecture?
- (RQ3) Which are the most suitable design processes that can be used to support the development of such an adaptive and adaptable architecture for ubiquitous learning?

3   State-of-the-Art Overview

In this section, we will present an overview of the "state of the art" of the most relevant topics that build the foundation of this chapter. We start by investigating the field of technology enhanced learning (TEL), which represents the application domain. Then, we present an overview over design-based research (DBR), the open architecture approach and conclude this section with the topic of domain-specific architecture. The interplay of those topics is the foundation of the presented work described later in this chapter.

3.1   Technology Enhanced Learning

The TEL-field has greatly benefited from advances in mobile, wireless, and sensor technologies along with web services and visualizations offered over the web [16]. Human and social aspects of computer system design, usage and evaluation are central issues in the TEL-field [17]. The aim of this field is to provide pedagogical and technological support in order to promote learning in different settings [18].

The latest developments in projects in the field of TEL show that a crucial aspect of ubiquitous learning activities is the mobility of the learners. Project like the one presented in [19–23] describe the technological support of learning activities with mobile devices. However, it is important to note that in all of those presented projects the context is slightly different. They offer support for learning activities but the way the methods are applied differ from each case and therefore findings from those works cannot easily be transported to new work. Either the use-cases differ too much from each other or the technical solution is specified for a single project and cannot be ported to another one. For instance, none of the presented projects are presenting a cross-platform solution for the supported mobile component in their project. The solutions are either based on iOS or Android, and therefore, it depends on the requirements if a solution is applicable.

Nonetheless, it is also possible to identify similarities in those projects. While combining projects in the field of TEL is possible to determine certain components that are present in the described projects. Those components are: (1) an authoring environment, (2) a software solution that supports learning activity directly, and (3) an analysis component. In these cases, an authoring environment allows to design learning activities. See for example [19] and [22]. However, the way this is conducted varies from the use-case. A trend in TEL is to move the ownership from a technical supported learning activity away from the developer to the teachers.

That means authoring environments need to be usable for nontechnical skilled users in order to allow them to design their own learning activities without being dependent on developers. The software solutions that offer support to perform a learning activity are focusing in the presented projects on mobile applications. As mentioned before, one aspect that gains immense attention recently in the field of TEL is the mobility of learners. Therefore, mobile applications like the one described in [23], are a natural step to support learning activities that are focusing on mobility.

However, a major challenge lies in the fact that the field of TEL is lacking a cross-platform mobile solution to meet the requirements of the fragmented marked. Analysis components like the one presented in [21] allow to reflect upon a performed learning activity by representing the data that is collected during an activity in a way that users can draw conclusions from it. An idea behind an analyses component is to enable students to easily perform visual exploration and analysis of data to facilitate the understanding of a particular phenomenon and to communicate findings and therefore widens the learning experience [24].

Reflecting upon our current knowledge and experiences from the field of TEL, two important issues can be identified for supporting ubiquitous learning activities:

1. A lot of work is conducted in the field of TEL but the existing solutions cannot adapt to new requirements or changing environments. Therefore, an adaptable solution is desired.
2. The same applies for the technological aspect of those solutions. They need to be adaptive to meet the rapidly changing technological environment when it comes to mobile devices.

3.2 Design-Based Research

Designing from development and user perspective is especially important when systems move away from a specific single use to an adaptive and adaptable perspective for multiple uses. The design challenge is also compounded in TEL because the diversity of the learners and the requirements for learning that differ from the professional use of systems. One approach that supports adaptivity is DBR that helps compose a coherent methodology that bridges theoretical research, development of technologies, and educational practice Collective.

This bridging is facilitated by the fact that the methods are grounded in the needs, constraints and interactions of local practice, ensuring to higher extent that research outputs have bearing on educational practices that include the development of architectures and systems. DBR envisions that researchers, practitioners and learners/users work together with the goal to produce or facilitate meaningful change in context of educational practices. As such, participatory design methods are frequently utilized in the field of TEL [25]. According to [26], the term DBR can be understood as encompassing a paradigm described by different terms in the literature including: design experiments [27]; design research [28, 29]; development research [30]; developmental research [31]. In this respect, DBR entails a series of approaches with the intent of producing new theories, practices, and artifacts that account for learning and teaching in educational practices.

The challenges with this expanded design approach are to find processes to enable the culture of critical, informed, and reflective design practices that includes a linguistic framework for communicating design knowledge [32]. It concerns developing a domain of practices along with established methods of resolving them. Mor's [32] approach is a triad of design narratives, scenarios, and patterns that aims for the social construction of design knowledge embedded in the design experiment.

Mor's trilogy combines the reflective practices of the narratives with observations of the scenarios to identify and formulate design patterns that can be shared. Together, these representations bridge the gap between theory and practice, mutually informing, and directing one another [25]. The use of different design approaches along with the different needs of the learners and teachers compared to professional users requires an adaptable approach for supporting ubiquitous learning across open architectures.

3.3 Open Architecture Approach

For domain experts in the TEL-field, we need to provide development insights, while integrating the technological resources and support necessary for successful implementation of the educational activities related to ubiquitous learning in authentic settings. In these environments, the real challenge is the need to match the dynamic requirements generated during the learning activities. To allow rapid technological changes to be smoothly reflected in everyday activities in this area, well-defined processes must exist to ensure the continual refinement of the applications and architectures developed. Thus, we need to provide solid foundations in terms of tackling the requirements necessary to support ubiquitous learning activities in a flexible manner. This flexibility, in terms of requirements, could, primarily, be enabled by utilizing the open architecture approach [33].

Multiple research efforts have occurred that have designed and developed software systems based on the notion of open architecture [33–43]. An open modular architecture was proposed by Byelozyorov et al. [35, 36] where open architecture combines several emerging and established technologies in order to provide tools for quickly developing prototypes of virtual worlds based on the web. The system is designed to be extensible, but also flexible due to the modular approach, which allows for the easy replacement of components. Another implementation of open architecture is based on a service-oriented approach specifically addressing the design of virtual organizations [37]. A cloud computing open architecture (CCOA) that integrates a service-oriented architecture is presented by Zhang and Zhou [39].

Cavuşoğlu et al. [38] present architectural details of an evolving open source/open architecture software framework in the field of medical informatics. Developing mobile location-based applications over the Internet can be also guided by the notion of open architecture as described by Jose et al. [41]. Their system uses a service-based approach in order to support location-based discovery. Furthermore, NASA (National Aeronautics and Space Administration, USA's space agency) uses an open architecture, component-based software tool for the development, integration, and deployment of mission operations software, with the challenge of integrating multiple applications into a single platform [42].

Oreizy [40] proposed a new software customization technique called open architecture software, which is a flexible approach to decentralized software evolution. His comparison framework was based on the concept of software open points, where independent third-party developers change a software system by modifying its architecture. One of the latest instantiations of a system based on open architecture is a project called mHealth [43].

The aim of mHealth is to create an open ecosystem of reusable, substitutable modules of basic functionalities, consisting of data exchange standards where existing and new systems would be interoperable [43]. The main concepts identified from the analysis of the efforts discussed above are presented in Table 1. One of the latest efforts in this direction was the work conducted by Vogel et al. [33, 34]. They have identified that the open architecture approach provides the means for establishing the design and development settings that capture the characteristics as attributes or constraints of a system.

As a result, the open architecture is characterized from three essential features that are instantiated into a collection of properties [21, 34]. The main enablers of the open architecture relevant for this chapter are:

- Flexibility, which provides solutions to be used by the users in a wide variety of settings and situations by easily addressing different user's requirements with minimum delays. Its measuring properties included: contextual, robustness, easiness, and cost-effectiveness.

Table 1 Open architectures—main concepts

Authors	Concepts
Oreizy [40]	Customization, open architecture software, flexible approach, software open points
Jose et al. [41]	Location-based, service-based, independent services, heterogeneity, openness in distributed systems
Cavuşoğlu et al. [38]	Open source/open architecture, developing APIs
Carrascosa et al. [37]	Service-oriented approach
Zhang and Zhou [39]	Cloud computing, service-oriented architecture
Lindsey [42]	Integrate multiple applications into a single platform
Estrin and Sim [43]	Reusable, substitutable modules, data exchange standards
Byelozyorov et al. [35]	Open modular architecture, prototypes, web, flexible, open system, open source code, open specification
Byelozyorov et al. [36]	Extensible
Vogel [33] and Vogel et al. [34]	Flexibility, customizability, extensibility, openness, open architecture, open source, open standards, interoperability, evolution, modularity, adaptability, service-oriented, constant changes, cost-effective

- Customizability, which allows users to easily customize features in the system and address their specific individual needs, usually without having access to the source code, thus reducing the deployment time. Its measuring properties included: specificity, easiness, and cost-effectiveness.
- Extensibility, which offers easy integration possibilities with other systems and/or tools that took into consideration future growth by expanding/enhancing the architecture with less costly upgrades. Its measuring properties included: adaptability, modularity, compatibility, easiness, and cost-effectiveness.

The above research projects are contributed by advancing the state of the art in the area of open architecture. Important aspects in this area are diversity and continued change of different development technologies and platforms. These projects have been deployed in multiple application domains, such as medical, governmental, military defense, space, social media, virtual worlds, and education. As such creates the need to explore the domain-Specific Architecture.

3.4 Domain-Specific Architecture

Building software from scratch over and over again in the form of ad hoc solutions in TEL is not feasible anymore. Ad hoc solutions bring one of the dominant challenge to engineers on reinventing the wheel [12]. Reusability is an important factor for improving quality and productivity [44]. Using a well-proven solution for a similar problem saves not only time during development but also ensures a high quality standard. Additionally, through reusing approved parts of software, it is possible to reduce the development costs and therefore save resources during the development process. To set a good foundation to create reusable software, it needs to be systematically planned and another crucial part is that the context of the domain needs to be considered, where software shares certain functionalities.

Domain aspects play a huge role during the development of reusable software components and are often not considered. Therefore, it is important to highlight this fact. Suganthy and Chithralekha [45] also stressing the importance of reusable software and are mentioning that it is necessary to engineer highly reusable software components from the beginning. They are claiming that DSSAs are a fitting solution to foster the creation of large software systems through defining components that have high reuse potential within a particular domain.

Indeed, during our last 5 years of development and research efforts we provided a number of solutions, where we gained critical knowledge and provided best practices by exploiting "common solutions to common problems" [12] for a particular domain and even subdomains. Domain-specific engineering can facilitate such processes when systematically planned and managed in the context of a particular domain [46]. Domain-specific software architecture (DSSA) is defined "as an assemblage of software components, specialized for a particular type of task (domain), generalized for effective use across that domain, composed in a standardized structure effective for building successful applications" [47].

Taylor et al. [12] state that engineers ought to focus on aspects of three inter-related areas while developing a DSSA: *domain, technology,* and *business.*

The *domain* area while independent from technology and business concerns, establishes a problem space. During this process, it defines for example but not exclusively: characteristics, domain vocabulary or motivation about why a domain exists.

The *technology* area consists of tools, applications, reusable components, infrastructure and methods that can be generally applied in the domain. Therefore, one could say this could be characterized as "solutions without problems."

The *business* area is mainly concerned with human goals: improving the quality in a certain domain for instance through the creation of new products, increasing the income or popularity. Those goals can motivate people to solve problems within a domain. However, this does not mean that domain-specific software needs to be sold or developed for the purpose of monetary profits; the goal can also be to optimize certain aspects of software engineering like reducing costs during the development process or improving the quality or service of products that are already developed.

Trasz [48] and Taylor et al. [12] state that a DSSA consists of at least two parts: a *domain model* and a *reference architecture.* Those are two crucial components of a DSSA, and therefore, a more detailed description is provided.

Domain Model. Architects often rely on their own knowledge to map domain-specific requirements onto generic software abstractions. Most of the time, this leads to the definition of repetitive tasks and architecture fragments, which can be particularly error prone. We therefore believe that architects need a more flexible approach to consider domain-specific aspects in their architectures. Software architects are rather tempted to manipulate domain-specific concepts, which could be seamlessly converted to technical artifacts and thus often leaving out nontechnical important domain-specific aspects.

According to Taylor et al. [12] the fundamental objective of a domain model is twofold: (1) "A domain model standardizes the given problem domain's terminology and its semantics. Together, the terminology and semantics are referred to as the domain's ontology." and (2) "A domain model provides the basis for standardized descriptions of problems to be solved in the domain." Therefore, the domain model is a crucial component in designing a DSSA since it provides the foundation for software that is developed in a certain domain.

Suganthy and Chithralekha [45] are mentioning in their work the importance of reusability issue and highlighting the role of a domain model in that case. They are arguing that the purpose of creating a domain model to develop domain artifacts that can be reused in other applications for a given domain. A domain model of activities for gathering and representing information on software that share common sets of capabilities and data. Thus, it is crucial to identify the reusable components in this step. According to Trasz [48] a domain model consists of sum of components that describes a domain and its assets. Those mentioned components are:

- Customer Needs
- Scenarios
- E/R Diagrams
- Dictionary
- Object Model
- Context Diagram
- Data-Flow Diagram
- State Trans Diagram

However, to design a domain model it is not necessarily needed to provide all of the mentioned components. It depends on the described domain and the importance a component has for the desired result. Therefore, a domain model can vary depending of the described domain. Figure 1 illustrates the described potential components of a domain model.

Reference Architecture. Martínez et al. [49] mention that the motivation behind reference architectures is to systematically reuse knowledge and software elements when developing concrete Software Architectures (SA) for new systems, to help with the evolution of a set of systems that stem from the same RA, or to ensure standardization and interoperability. However, although the adoption of an RA might have plenty of benefits for an organization, it also implies several challenges, among them the need for an initial investment. An RA encompasses the knowledge about how to design concrete SA of systems of a given application domain.

According to [50], a reference architecture is a template solution for a concrete architecture for a certain domain. It is a software architecture where structures, relations, and elements provide solutions for either a particular domain or a group of software systems. Often, it consists of a list of functions and their interfaces can be defined at a high abstraction level and a common vocabulary to discuss

Fig. 1 Potential components
of a domain model

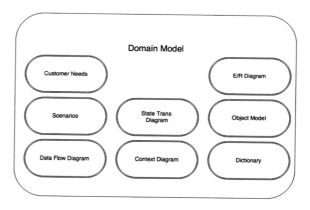

implementations is attached to highlight the commonality of solutions. Since, creating a reference architecture is a complex process, it is usually an iterative and community-driven mechanism.

Additionally, Martinez et al. [51] are mentioning the fact that reference architectures increasing speed, reduce operational expenses, and improve quality in software systems. Furthermore, they conducted a literature survey and pointing out the benefits and drawbacks of reference architectures gathered from the literature. The list of identified benefits they provide reads as follows (taken from [51]):

- Standardization of concrete architectures of systems.
- Facilitation of the design of concrete architectures for system development and evolution, improving the productivity of system developers.
- Systematic reuse of common functionalities and configurations in systems generation, implying shorter time-to-market and reduced cost.
- Risk reduction through the use of proven and partly prequalified architectural elements.
- Better quality by facilitating the achievement of quality attributes.
- Interoperability of different systems.
- Creation of a knowledge repository that allows knowledge transfer.
- Flexibility and a lower risk in the choice of multiple suppliers.
- Elaboration of the organization mission, vision, and strategy.

The list of identified drawbacks from the same literature survey is significantly shorter and reads as follows (taken from [51]):

- The need for an initial investment.
- Inefficient instantiation for the organization's systems.

This clearly expresses the advantages the existence of a reference architecture has when it comes to developing new applications with a similar background. Furthermore, the authors who performed the literature survey present an own study to extend the list of benefits and drawbacks taken from the literature. The study consisted of nine projects in diverse organizations and covered three different stakeholders: software architects, architect developers and application builder. The founding of this study is an enhanced list of benefits, and drawbacks. The benefits read as follows (taken from [51]):

- Reduced development costs.
- Reduced maintenance costs.
- Easier development and increased productivity of application builders by architecturally significant requirements already addressed and RA artifacts.
- Incorporation of latest technologies, which among other things facilitates the recruitment of professionals with the required technological skills
- Applications more aligned with business needs.
- Homogenization (or standardization) of the development and maintenance of a family of applications by defining procedures and methodologies.
- Increased reliability of RAs software elements, which are common for applications, which have been tested and matured, with the reliability that it implies.
- The consulting company harvests experience for prospective RA projects.
- Reusing architectural knowledge can speed up prospective RA projects and reduce time-to-market.
- They gain reputation for prospective client organizations and gain organizational competence.
- Previous experience reduces the risks in future projects because a "to-be" model exists. It can be used in projects without very specific requirements.
- It provides a shared architectural mindset.
- It turns tacit knowledge into explicit knowledge in the reference model. Some tool support (e.g., wiki technologies) helps in managing such knowledge.

And the drawbacks from the mentioned study read as follows (taken from [51]):

- Additional high or medium learning curve for using the RA features.
- Limited innovation by giving prescriptive guidelines for applications.
- Applications' dependency over the RA. When applications have requirements that the RA does not offer yet, applications development is stopped.
- Complexity. Participants who indicated that the use of the RA is complex.
- None. Responders who indicated that RA adoption presents no drawbacks.
- Wrong decisions about the technologies to be used in all the applications.

That underlines even stronger the importance and advantages that a reference architecture brings when it comes to develop new software.

3.5 Summary

We provided an overview about the state of the art in the TEL field, open architecture, DBR and domain-specific architecture. It was shown that within the TEL field, technologies, and learning environments are evolving at a fast pace. This opens opportunities for developers and researchers but also brings challenges.

Existing approaches are not built to keep up with the fast pace of development and it is hard to identify common patterns that allow to reuse certain components of existing approaches.

4   Technical Motivation

In the following section, we present two of our projects that serve as a motivation for our current work.

4.1   GEM

GEM (Geometry Mobile) is a project that was initiated as part of education in mathematics [52]. The purpose of GEM is to clarify the concept of triangulation by using mobile devices in a field. The whole activity consists of finding certain points using the GPS sensor of mobile devices. In preparation of the activity, six fixed points, so-called landmarks, are assigned that are placed within a fixed form of a rectangle.

The goal of one task is to find a point that has a certain predefined distance from two of the fixed points. The architectural approach in this project was based on different components and features with some instantiated central concepts Resource and Connector. These concepts are instantiated because of their reusability benefits in terms of flexibly adding more recourses and even new applications into the system. Moreover, the notion of Agent in this approach mainly deals with users and how they are modeled; the Device concept provides a meta-model of different devices available. Whereas Rule and Logic concepts mainly deal with system orchestration that provide additional controls and features into the system [53]. In general, this solution utilizes the concepts of self-adaptation that makes the architectural approach resilient to uncertainties in runtime environments.

4.2   LETS GO

One of the aims Learning Ecology through Science with Global Outcomes (LETS GO) research project was to support open inquiry learning using mobile science collaborators to provide open software tools and resources and participation frameworks for user project collaboration, mobile data and media capture, publishing, analysis, and reflection. From a software engineering perspective, fulfilling these aims introduced a number of stakeholders as well as functional and architectural requirements [54, 55].

Fig. 2 The LETS GO solution in use: mobile data collection and data visualization

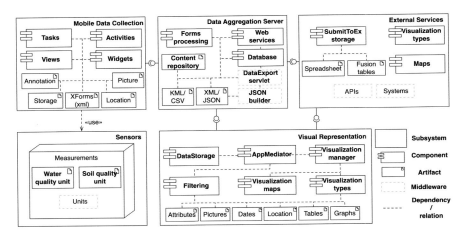

Fig. 3 LETS GO architectural artifacts and components

The overall evolution of our LETS GO case solution was informed by tests that we conducted with more than 500 users over a 4-year period (see Fig. 2). The end users' (teachers and students) feedback was an important and continuous input throughout the software development lifecycle.

LETS GO system and its architectural approach. The software system that we developed through the iterative development cycles during the last 4 years has resulted in a modular and layered architecture, based on a service-oriented approach that utilizes open-source components, open data standards, and usable formats. The software system has gone through an evolutionary prototyping approach to become a stable and robust platform for mobile data collection, aggregation, and data visualization.

Figure 3 presents an overview of the architecture resources of the LETS GO that are organized into different subsystems that integrate sensors, mobile units, server side components, and visualization and external services.

The mobile data collection subsystem shows some of the main components (tasks, activities, views and widgets) that serve the data collection. This subsystem utilizes an open standard data format, which flexibly renders it in order to dynamically reflect upon the user requirements. The process of designing these forms was performed usually with a simple text editor or the authoring tool and uploaded to the data aggregation subsystem.

The data aggregation subsystem takes care of validating, managing, and processing these forms to the mobile clients. Here, all of the forms and their related data collected through the mobile clients are uploaded and saved. Some of the main components in this subsystem are related to the forms processing, web services, databases, and open data standards used for increasing extensibility. The external services subsystem contains components that utilize diverse open web APIs.

The Visual Representation subsystem is controlled by one of the central components, AppMediator, which handles data storage, different visualization techniques, and filtering functionalities. All of the visualization techniques are integrated by using available open web APIs. As for the sensors, the system utilizes and makes use of some of the basic units for measurement purposes related to water or soil quality.

4.3   Conclusion

Those two projects are good examples for the fact that, although they have a similar sequence of actions, they were built on two vastly different approaches. This supports our goal to define a foundation for software solutions within ubiquitous learning that is able to adapt to the dynamic aspects in this field and at the same time does not require a completely new software solution but it is able to reuse components. Therefore, in the following, we are proposing a DSA that can serve as a foundation to software solutions that follow a similar course of actions.

5   Toward a Domain-Specific Architecture for TEL Software

The main driving force of this chapter is to provide an integrated approach toward a need for a domain-specific architecture in TEL and beyond. Thus, the time has come to deal with such integration. Table 2 lists initial concepts classified into domain semantics, concepts, and realization identified through the state-of-the art-study conducted above (see Sect. 3). In order to ensure that functional, design, and domain concerns are managed well into the development of our technological solutions, Table 2 sets the scene toward addressing the following challenges:

1. Providing better understanding on how to translate domain goals into technology and design areas,
2. Better explaining technology constraints to the specific domains and even subdomains, and
3. Properly communicating design ideas to the technology and domain areas.

Table 2 Domain semantics, concepts and realization

Domain semantics	Concepts	Realization
Ubiquitous learning	Openness	Identify domain needs
Physical and environmental	Open architecture	Capture domain assets
Changing settings	Open source	Analyze domain assets
Across context	Open standards	Design domain model
Learning in context	Flexibility	Specify technology
Location	Customizability	Use reusable components
Time	Extensibility	Use standardization methods;
Activities	Interoperability	Define components
Preferences	Evolution	Define relations
Data collection	Modularity	Design reference architecture
Reflection	Adaptability	
Analysis	Service-oriented	
Exploration	Reusability	
Collaboration	Cost-effective	
Mobility		
Interactivity		
Engagement		
Curiosity		
Flexibility		

In order to further decompose the ideas presented in Table 2 that could lead toward a domain-specific Architecture there is a need to define a domain model and based on the domain knowledge provide a potential reference architecture that then serves as a template for software development in the described domain.

As described in our state-of-the-art study (see Sect. 3), the design processes of DBR also can be seen to be parallel and informed the development of domain-specific architecture for the TEL-field. DBR focuses on understanding the needs of the customers (i.e., learners and teachers), the complex scenarios of learning across different scenarios (fields of study and types of education). Designing for learning needs to be seen as different to other design activities since the learners represent a different type of stakeholder. Inspired by Soloway and colleagues [56] early work, learner-centered design (LCD) and Norman and Draper's [57] user-centered design.

Our work takes the view that design for learning needs to take into consideration, the conceptual distance between the learner and the computer (technological system). Our learner-centered approach takes into consideration the goals of the users and the result of using the tool (computer system) while keeping in mind the conceptual distance, "the gulf of expertise" that lies between the novice and the developed understanding or expertise embodied by a more capable peer.

Fig. 4 Factors that influence
the DSA

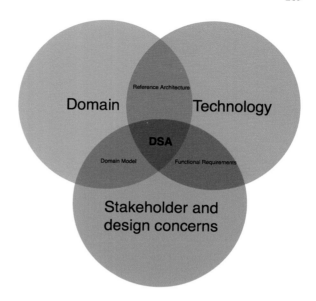

Figure 4 shows what factors influence the design of a DSA. As shown, domain, technology, and stakeholders are the driving forces when it comes to design a DSA. The process of designing the DSA is an iterative process. We start by providing an initial overview of a domain model. From there, an initial model of a reference architecture is presented.

5.1 Domain Model

This first design of the domain model and the reference architecture by all means are not the final outcome that will lead to the domain-specific architecture but are the first iterations of many that eventually will lead to a final domain-specific architecture. Figure 5 displays this process.

As described in Fig. 5 the main concepts during the domain engineering that will lead to a domain model are (1) identifying domain assets, (2) capturing domain assets, and (3) analyze domain assets. Those steps will lead to the first iteration of the domain model.

As mentioned before, a domain model may consist of numerous components. To perform the mentioned steps, we will provide initial realizations of relevant components within the TEL domain. An overview of potential elements of a domain model is shown in Fig. 1. In the following, we will provide initial diagrams that will serve as an initial approach to define a domain model.

Due to the iterative process, these diagrams may go through severe changes in the future but the presented diagrams provide a first attempt to design a domain model. We present a context diagram to clarify the scenarios that we are trying to cover (see Fig. 6).

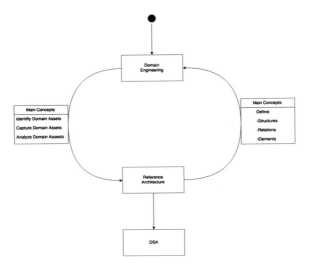

Fig. 5 The process for designing a domain-specific architecture

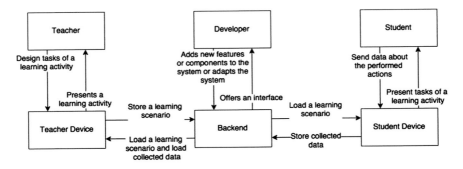

Fig. 6 Context diagram for the TEL domain

Those scenarios are derived from the state-of-the-art study that was conducted before (Sect. 3). The study shows that three main user need to be considered, the teachers that are preparing and supervising a learning activity, the students who perform the learning activity and developers that are constantly working at ICT (Information and Communications Technologies) solution to support teachers and students (see Fig. 7).

Also, the state-of-the-art study shows that a learning activity can be divided into three main activities. A learning activity can be:

1. An activity that requires the user to collect data with sensors (internal or external), i.e., taking pictures with an in-build camera or measure water quality with external sensors.

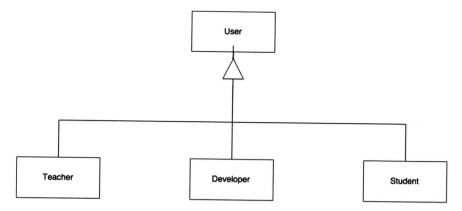

Fig. 7 User type taxonomy

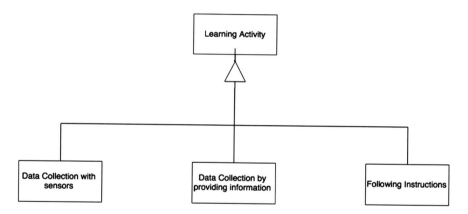

Fig. 8 Learning activity taxonomy

2. Collect data by providing information, i.e., answering questions of a quiz.
3. Following instructions. Figure 8 illustrates this.

Matching the context diagram (see Fig. 6) that shows the potential application scenarios, the data flow of software in educational settings is presented in Fig. 9. Teachers either create or edit activities and those activities are stored in a data storage. Students get a list of activities that they can perform. The data that is generated during these activities is also stored in the data storage. This data can then be visualized for teachers to analyze the conducted activity.

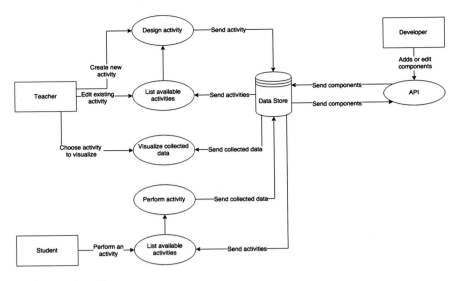

Fig. 9 Data flow diagram

5.2 Reference Architecture

Based on the described domain model and the state-of-the-art study are in Sect. 3, we present a first attempt to design a reference architecture. The reference model describes the components that a system should posses to match the requirements gathered from the domain model and the state-of-the-art study. Furthermore, our proposed reference architecture suggests defines technologies to fulfill the description of the Domain Model and the state-of-the-art study.

As discussed above in Sect. 3, one major challenge in the field of TEL is the possibility for end users with no technical background to easily design and deploy their own mobile applications. Thus, we are proposing an architecture with three main components (see Fig. 10 and a back-end system. The three components are: authoring component, a mobile application, and visualization component. Those components result from the domain model described before. They cover the use-cases and the presented learning activities. The back-end system handles functionalities like data management, logging, user management, and more.

Another major challenge is providing cross-platform mobile applications, which can be executed on a mobile web-browser to meet the challenges of the fragmented mobile device market. The suggested technologies are based on the state-of-the-art study described in Sect. 3 and are mainly a result based on the findings from open architecture. We are suggesting using complete web-based technologies to increase the level of adaptive and adaptable factors. Web technologies and the fact that the components are loosely coupled make it easy for developers to add new feature or change existing ones. Baloian et al. argued that HTML5 has an enormous potential

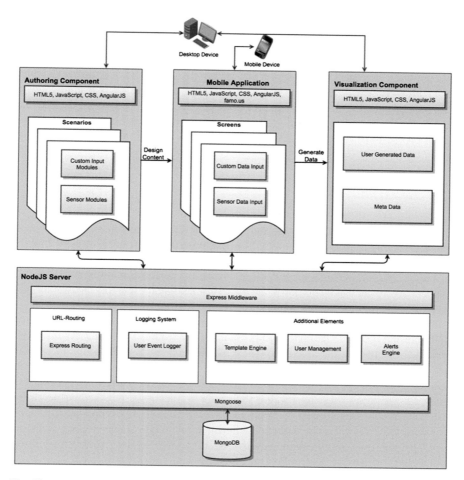

Fig. 10 Toward a reference architecture design

for executing learning scenarios with mobile devices [58]. Nonetheless, they are focusing on learning scenarios but not on authoring, deployment, or analysis.

In our architecture, the whole communication between the components functions over an API so that it is easy to access the data from outside. Additionally, the usage of APIs and loosely coupled components allows the reusage of certain components in other scenarios if needed. If for instance, only the authoring component is needed in a project, it can be easily integrated using APIs.

6 Implementation and Deployments

Based on our domain model and initial reference architecture we developed the mLearn4web system. Coming from the lessons learned from the LETS GO and GEM projects, we designed a system that is able to design mobile applications, automatically deploy them and offer an initial analyze of the generated data by the mobile application. This system is designed to not only cover the functionalities of LETS GO and GEM but also more scenarios with a unified architecture behind it.

6.1 mLearn4web

mLearn4web is a system that tackles the challenge of the introduction of new technologies in educational settings. The immense distribution of modern mobile devices for students is an opportunity, which is more and more recognized nowadays. Mobile devices are no longer banned to perform learning activities but teachers try to take advantage of the fact that nearly all students posses mobile devices like smart phones or tablets.

In fact, some countries like Sweden changed their national curriculum to force the inclusion of ICT in educational settings [59]. Thus, in order to take full advantage of the available technological setup—the huge availability of mobile devices—it is necessary to provide support for teachers to compose their own learning applications for those devices. However, teachers usually do not possess the skills and knowledge to develop their own applications that fit their needs so that teachers can make use of the setup to its full potential. Therefore, we introduce a system that allows users without technical knowledge or developing skills to develop their own mobile applications [60, 61]. This system consists of three components: authoring tool, mobile application, and visualization tool.

The authoring tool allows the design of mobile application with simple, well-known interaction methods like drag and drop. Teachers can design their mobile applications based on designing screens. In the authoring tool, it is possible to create a number of screens, rearrange their order and delete them. Onto those screens, it is possible to add content through drag and drop predefined elements onto the screen.

The resulting mobile application is automatically deployed with the content that was specified before. The order of the screen, defined in the authoring tool with its content is displayed on a mobile device. Some modules allow use of internal sensors of modern mobile devices like camera, microphone, or GPS. Thus, it is possible to use the designed mobile applications for data collection purposes.

The data that is collected during the usage of the mobile application is presented in a visualization tool. In this tool, data that belong to the same screen are considered to be in context to each other. For instance, if on one screen a picture was taken and on the same screen the location was also stored, it is assumed that those

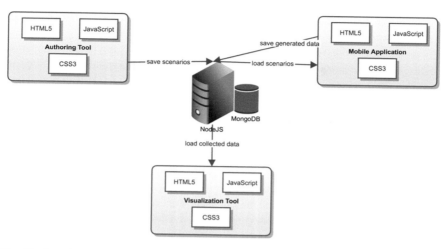

Fig. 11 Overview about the mLearn4web system

data are in context to each other. Therefore, the data is presented in an appropriate way. In this case, a map is presented where markers indicate that additional data is available. When clicking on the marker the picture is presented on the map.

The system was evaluated during workshop with teachers that had no technical background at all [62, 63]. As a result it was concluded that it was no problem for the teacher to design and deploy learning scenarios that they envisioned despite the lack of technical background and knowledge. Especially, noticeable is the fact that teachers learned extremely fast to use the system even though they struggled with certain functionalities in the beginning. That underlines our hypothesis that the system is a good solution that allows users that do not have programming skills to design their own mobile application fitting to their needs. Figure 11 provides an overview about the mLearn4web system and its components.

Today, the system is available online and is used by schools where we introduced the system and explained it to teachers. However, we are noticing activities from schools where we did not provide any help or introduction to the system, which indicates that this system can be used without training the teachers to use it.

6.2 PELARS

The Practice-based Experiential Learning Analytics Research And Support (PELARS) research project aims to generate, analyze, use, and provide feedback for analytics derived from hands-on project-based learning activities [64]. The overall research aim of the PELARS project can be summarized as: How can physical learning environments and the use of hands-on digital fabrication technologies are better designed for ambient and active data collection for analytics that can better support learning?

The project addresses three different learning contexts (university interaction design, engineering courses, and high school science) across multiple settings in Europe. The goals of the project are first to define learning (skills, knowledge, competencies) that is developing, and how we can assess it in the frame of learning analytics in the workshop setting. The PELARS project has developed an intelligent system for collecting activity data (moving image-based and embedded sensing) for diverse learning analytics (data-mining, reasoning, visualization) with active user-generated material (mobile and web) from practice-based activities.

A core part of the PELARS system is the learners' self-documentation for planning, documenting, and reflecting on their work. We applied mLearn4web "system" to develop the personal mobile and web tool for documentation. The personal mobile tool is a web-based system that allows the learners to send rich media and text as a live stream of data into the PELARS learning analytics system. The students access it from their personal mobile devices and laptops.

6.3   TriGO

TriGO [65] is a successor of the GEM project [52]. One big challenge that got addressed in TriGO is related to the preparation of an activity. In GEM, a developer was needed to define the exact locations of the landmarks in a XML-file and copy those XML-files to the mobile devices that were used during the activity. This issue was solved by providing an authoring tool that allows users to place a rectangle on a map and therefore assign the six landmarks to its locations.

This issue was solved by re-building the application according to the presented DSA. The technological setup of the authoring tool of mLearn4web was used but due to the fact that it was implemented in an adaptive and adaptable manner, it was easy to adapt the authoring tool to the new requirements of the TriGO project.

Each activity is assigned an ID and a password that needs to be entered in the TriGO mobile application. After this is provided the activity that belongs to the ID and password combination is available in the mobile application.

With the addition of the authoring tool, it is no longer necessary for a developer to prepare an activity but teachers can do it by themselves and thus, it is easier to provide this project to a big number of schools.

Furthermore, according to the described DSA, a visualization tool was also implemented. After an activity is performed, it is possible to upload all data that includes all tries to find a certain point. This data can then be presented to analyze the strategies that students chose to find a certain location. Such a strategy can be to find a spot first that matches one of the desired distances and go in a circle toward the second landmark until the desired spot is reached. This whole process, for example, can be visualized by limiting the visualization to a certain group and a certain task. The whole process is illustrated in Fig. 12.

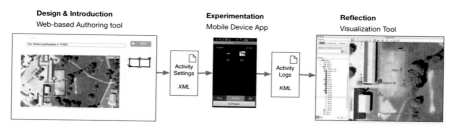

Fig. 12 The complete process of TriGO

The concept of the activities was evaluated extensively in recent years [52]. However, the addition of the authoring tool was not included in those evaluations. Therefore, a workshop with ten teachers was conducted and as a result it was indicated that it was possible for the teachers to define the locations where the activity should take place by their own.

7 Conclusions and Outlook

In this chapter, we discussed that the field of ubiquitous learning is constantly evolving due to the rapid developments in technologies and learning environments. This leads to numerous variants of implementations for ubiquitous learning software. Often, these solutions have very specific limitations. We showed that in most cases the support for mobile devices is limited to a certain device and/or operating system, although a cross-platform solution is desired in most cases.

Another issue that arises with a wide spread of diverse solutions is that these solutions are not compatible to each other, and therefore, it is usually not possible to reuse a solution for a new or other projects. This repetitive cycle allows an endless stream of different solutions to emerge in the field of TEL.

Our literature study revealed three common components that are generally used in ubiquitous learning software solutions. These components are: (1) authoring environment, (2) an application to support performing a learning activity, and (3) an analyzing tool.

We have proposed that adapting an open architecture approach supported by DBR can help providing continuity to ubiquitous learning. Our intention is to provide a better understanding how to translate domain goals into designing technology. This intention can explain technology constraints across specific domains and properly communicate design ideas between stakeholders for technology and the domain areas.

The core contribution of the chapters is the initial DSA for the field of TEL that offers a general foundation for ubiquitous learning software. The presented DSA consists of the three components (authoring, activities, and analysis) to cover the broad landscape of ubiquitous learning software. Those components are loosely coupled, so that it is easy to replace and/or remove one of the components if necessary.

By describing the first steps toward the DSA, we are also tackling the research questions posed earlier in this chapter. The presented initial model of a DSA is the first step toward an answer to RQ1. We are showing that this architecture can be easily adapted to evolving requirements by presenting the use-cases where the architecture was applied. The fact, that we are strongly considering domain aspects in our DSA allows us to identify the fast changing and dynamic conditions in the field of ubiquitous learning. Therefore, presenting a DSA and thus, involving domain aspects is a solution to the RQ2. During the development of the DSA, we realized that applying DBR is a fitting process to identify requirements for domain aspects and also requirements for an adaptive and adaptable architecture and thus answering RQ3.

During the course of this work, we realized that proposing a DSA is a highly iterative process and therefore we are not offering a final version of a DSA but a first iteration that has space to evolve and adapt to new requirements. Therefore, this chapter represents only the first steps toward a domain-specific architecture for the field of TEL. This architecture aims to meet challenges that arise due to the rapid development in the field. Not only technologies but also the environment is constantly evolving and therefore there is a need for an adaptive and adaptable approach like the one presented in this chapter.

Next Steps. In next steps the domain model as well as the Reference Architecture will be refined. Additionally, we will apply the proposed DSA to more projects to get more data for a more detailed evaluation. The concept of the proposed DSA is not limited to the field of TEL. Since we are focusing our work on adaptive and adaptable elements in our DSA, it is easily possible to transport many concepts to other fields. In addition to projects from the TEL domain, the concept of our DSA can be applied in other settings.

One example is the field of smart housing. The idea is that users can design their own scenarios with the sensors that are available in smart housing. Using an authoring environment, specifying the actions that are desired. For instance, activating the heating system when the temperature is below a certain temperature. A mobile application can be used to control certain actions and a visualization tool to monitor all the actions. Thus, this sort of system can be build with our proposed DSA although it is outside the TEL domain and the Domain Model needs to be adjusted.

The same can be applied within the medical field. Doctors can create activities for patients that they need to perform with mobile devices. For example, tracking some fitness tests with a mobile device. The results can then be analyzed in a visualization environment. Again, the domain model would need to be adjusted to the application field.

References

1. Wu, T.-T., Yang, T.-C., Hwang, G.-J., Chu, H.-C.: Conducting situated learning in a context-aware ubiquitous learning environment. In: Fifth IEEE International Conference on Wireless, Mobile, and Ubiquitous Technology in Education, 2008. WMUTE, pp. 82–86 (2008)
2. Chang, B., Wang, H.Y., Lin, Y.S.: Enhancement of mobile learning using wireless sensor network. IEEE Learn. Technol. Newslett. **11**, 22–25 (2009)
3. Kohen-Vacs, D., Kurti, A., Milrad, M., Ronen, M.: Systems integration challenges for supporting cross context collaborative pedagogical scenarios. In: Collaboration and Technology, pp. 184–191. Springer, Berlin (2012)
4. Kukulska-Hulme, A., Sharples, M., Milrad, M., Arnedillo-Sánchez, I., Vavoula, G.: Innovation in mobile learning: a European perspective. Int. J. Mob. Blended Learn. **1**, 13–35 (2009)
5. Breivold, H.P., Crnkovic, I., Larsson, M.: Software architecture evolution through evolvability analysis. J. Syst. Softw. **85**, 2574–2592 (2012)
6. Yu, L., Ramaswamy, S., Bush, J.: Symbiosis and software evolvability. IT Prof. **10**, 56–62 (2008)
7. Specht, M.: Design of contextualised mobile learning applications. Increasing Access **61** (2014)
8. Björgvinsson, E., Ehn, P., Hillgren, P.-A.: Design things and design thinking: contemporary participatory design challenges. Des. Issues **28**, 101–116 (2012)
9. Wasserman, A.I.: Technical and Business Challenges for Mobile Application Developers, http://mobicase.org/2011/archive/2010/docs/Wasserman_mobicase2010.pdf (2010)
10. Wasserman, A.I.: Software engineering issues for mobile application development. In: Proceedings of the FSE/SDP workshop on Future of software engineering research, pp. 397–400 (2010)
11. Taivalsaari, A., Mikkonen, T.: Objects in the cloud may be closer than they appear towards a taxonomy of web-based software. In: 2011 13th IEEE International Symposium on Web Systems Evolution (WSE), pp. 59–64 (2011)
12. Taylor, R.N., Medvidovic, N., Dashofy, E.M.: Software Architecture: Foundations, Theory, and Practice. Wiley Publishing (2009)
13. Lew, P., Olsina, L.: Relating user experience with mobileapp quality evaluation and design. In: Current Trends in Web Engineering, pp. 253–268. Springer, Berlin (2013)
14. Sotsenko, A., Jansen, M., Milrad, M.: Using a rich context model for a news recommender system for mobile users. In: Proceedings of 2nd International Workshop on News Recommendation and Analytics (2014)
15. Wong, S., Sun, J., Warren, I., Sun, J.: A scalable approach to multi-style architectural modeling and verification. In: 13th IEEE International Conference onEngineering of Complex Computer Systems, 2008. ICECCS 2008, pp. 25–34 (2008)
16. Winters, N., Price, S.: Mobile HCI and the learning context: an exploration. In: Proceedings of Context in Mobile HCI Workshop at MobileHCI05 (2005)
17. Tchounikine, P.: Computer Science and Educational Software Design: A Resource for Multidisciplinary Work in Technology Enhanced Learning/Pierre Tchounikine
18. Kurti, A.: Exploring the Multiple Dimensions of Context: Implications for the Design and Development of Innovative Technology-Enhanced Learning Environments (2009)
19. Mulholland, P., Anastopoulou, S., Collins, T., Feisst, M., Gaved, M., Kerawalla, L., Paxton, M., Scanlon, E., Sharples, M., Wright, M.: nQuire: technological support for personal inquiry learning. IEEE Trans. Learn. Technol. **5**, 157–169 (2012)
20. Hwang, G.-J., Tsai, C.-C., Chen, C.Y., et al.: A context-aware ubiquitous learning approach to conducting scientific inquiry activities in a science park. Australas. J. Educ. Technol. **28**, 931–947 (2012)

21. Vogel, B., Kurti, A., Milrad, M., Johansson, E., Müller, M.: Mobile inquiry learning in Sweden: development insights on interoperability, extensibility and sustainability of the LETS GO software system. Educ. Technol. Soc. **4522**, 43–57 (2014)
22. Giemza, A., Bollen, L., Hoppe, H.U.: LEMONADE: field-trip authoring and classroom reporting for integrated mobile learning scenarios with intelligent agent support. Int. J. Mob. Learn. Organ. **5**, 96–114 (2011)
23. Kim, S., Mankoff, J., Paulos, E.: Sensr: evaluating a flexible framework for authoring mobile data-collection tools for citizen science. In: Proceedings of the 2013 conference on Computer Supported Cooperative Work (CSCW'13), pp. 1453–1462 (2013)
24. Heer, J., Viégas, F.B., Wattenberg, M.: Voyagers and voyeurs: supporting asynchronous collaborative information visualization. In: Proceedings of the SIGCHI conference on Human factors in computing systems, pp. 1029–1038 (2007)
25. Mor, Y., Winters, N.: Design approaches in technology-enhanced learning. Interact. Learn. Environ. **15**, 61–75 (2007)
26. Wang, F., Hannafin, M.J.: Design-based research and technology-enhanced learning environments. Educ. Technol. Res. Dev. **53**, 5–23 (2005)
27. Brown, A.: Design experiments: theoretical and methodological challenges in creating complex interventions in classroom settings. J. Learn. Sci. **2**, 141–178 (1992)
28. Cobb, P., Confrey, J., DiSessa, A., Lehrer, R., Schauble, L.: Design experiments in educational research. Educ. Res. **32**, 9–13 (2003)
29. Bielaczyc, K.: Design Research: Theoretical and Methodological Issues. Allan Collins Northwestern University Diana Joseph University of Chicago
30. Van den Akker, J.: Design Methodology and Developmental Research in Education and Training. Presented at the (1999)
31. Richey, R.C., Klein, J.D.: Developmental research methods: creating knowledge from instructional design and development practice. J. Comput. High, Educ (2005)
32. Mor, Y.: SNaP! Re-using, sharing and communicating designs and design knowledge using scenarios, narratives and patterns. In: Handbook of Design in Educational Technology. London, pp. 1–12 (2014)
33. Vogel, B.: An Open Architecture Approach for the Design and Development of Web and Mobile Software (2014)
34. Vogel, B., Kurti, A., Mikkonen, T., Milrad, M.: Towards an open architecture model for web and mobile software: characteristics and validity properties. In: Computer Software and Applications Conference (COMPSAC), 2014 IEEE 38th Annual, pp. 476–485 (2014)
35. Byelozyorov, S., Pegoraro, V., Slusallek, P.: An open modular architecture for effective integration of virtual worlds in the web. In: International Conference on Cyberworld, pp. 46–53 (2011)
36. Byelozyorov, S., Rubinstein, D., Pegoraro, V., Slusallek, P.: An open modular middleware for interoperable virtual environments. In: International Conference on Cyberworlds (CW), pp. 94–100 (2013)
37. Carrascosa, C., Giret, A., Julian, V., Rebollo, M., Argente, E., Botti, V.: Service oriented MAS: an open architecture. In: Decker, S., Sichman, A., Sierra, B., Castelfranchi, G. (eds.) Proceedings of 8th International Conference on Autonomous Agents and Multiagent Systems (AAMAS 2009), pp. 1291–1292. International Foundation for Autonomous Agents and Multiagent Systems, Budapest, Hungary (2009)
38. Cavusoglu, M.C., Goktekin, T.G., Tendick, F.: Gi{PS}i: a framework for open source/open architecture software development for organ-level surgical simulation. IEEE Trans. Inf. Technol. Biomed. **10**, 312–322 (2006)
39. Zhang, L.J., Zhou, Q.: {CCOA}: Cloud computing open architecture. In: IEEE International Conference on Web Services (ICWS 2009), pp. 607–616 (2009)
40. Oreizy, P.: Open architecture software: a flexible approach to decentralized software evolution (2000)

41. Jose, R., Moreira, A., Meneses, F., Coulson, G.: An open architecture for developing mobile location-based applications over the Internet. In: Proceedings of the Sixth IEEE Symposium on Computers and Communications, pp. 500–505 (2001)
42. Lindsey, A.E.: Component-based tool for mission operations software deployment. In: Computational Sciences Division. NASA Ames Research Center, pp. 1–8. American Institute of Aeronautics and Astronautics (2009)
43. Estrin, D., Sim, I.: Open mHealth architecture: an engine for health care innovation. Science (80-) **330**, 759–760 (2010)
44. Basili, V.R., Briand, L.C., Melo, W.L.: How reuse influences productivity in object-oriented systems. Commun. ACM **39**, 104–116 (1996)
45. Suganthy, A., Chithralekha, T.: Domain-specific architecture for software agents. J. Object Technol. **7**, 77–100 (2008)
46. De Almeida, E.S., Alvaro, A., Garcia, V.C., Nascimento, L., Meira, S.L., Lucrédio, D.: Designing domain-specific software architecture (DSSA): towards a new approach. In: Working IEEE/IFIP Conference Software Architecture 2007, 0–3 (2007)
47. Hayes-Roth, F.: Architecture-based acquisition and development of software: guidelines and recommendations from the ARPA domain-specific software architecture (DSSA) program. Teknowledge Fed. Syst. Version. **1** (1994)
48. Tracz, W.: DSSA (domain-specific software architecture): pedagogical example. ACM SIGSOFT Softw. Eng. Notes. **20**, 49–62 (1995)
49. Martínez Fernández, S.J., Ayala Martínez, C.P., Franch Gutiérrez, J., et al.: A reuse-based economic model for software reference architectures (2012)
50. Wilson, A., Lindholm, D.M., LASP, C.U.: Towards a domain specific software architecture for scientific data distribution. In: AGU Fall Meeting Abstracts, p. 1609 (2011)
51. Martínez-Fernández, S., Ayala, C.P., Franch, X., Marques, H.M.: Benefits and drawbacks of reference architectures. In: Software Architecture, pp. 307–310. Springer, Berlin (2013)
52. Gil de la Iglesia, D., Calderon, J.F., Weyns, D., Milrad, M., Nussbaum, M.: A Self-adaptive multi-agent system approach for collaborative mobile learning. IEEE Trans. Learn. Technol. **99**, 1 (2015)
53. Pettersson, O., Gil de la Iglesia, D.: On the issue of reusability and adaptability in M-learning systems. In: 6th IEEE International Conference on Wireless, Mobile and Ubiquitous Technologies in Education (WMUTE), pp. 161–165 (2010)
54. Vogel, B., Kurti, A., Spikol, D., Milrad, M.: Exploring the benefits of open standard initiatives for supporting inquiry-based science learning. In: Wolpers, M., Kirschner, P.A., Scheffel, M., Lindstaedt, S.E., Dimitrova, V. (eds.) Sustaining TEL: From Innovation to Learning and Practice, pp. 596–601. Springer, Berlin Heidelberg (2010)
55. Vogel, B., Spikol, D., Kurti, A., Milrad, M.: Integrating mobile, web and sensory technologies to support inquiry-based science learning. In: 6th IEEE International Conference on Wireless, Mobile and Ubiquitous Technologies in Education (WMUTE 2010), pp. 65–72 (2010)
56. Soloway, E., Guzdial, M., Hay, K.E.: Learner-centered design: the challenge for HCI in the 21st century. Interactions. **1** (1994)
57. Norman, D.A., Draper, S.W.: User centered system design: new perspectives on human-computer interaction. Erlbaum, Hillsdale, N.J. (1986)
58. Baloian, N., Frez, J., Diego, E.-U., Jansen, M., Zurita, G.: The future role of HTML5 in mobile situated learning scenarios. In: Yu, S. (ed.) 10th World Conference on Mobile and Contextual Learning (mLearn), pp. 249–257. Normal University of Beijing, Beijing, China (2011)
59. The Swedish National Agency for Education (Skolverket).: Curriculum for the Compulsory School, Preschool Class and the Leisure-Time Centre 2011. Skolverket, Stockholm (2011)
60. Zbick, J.: A web-based approach for designing and deploying flexible learning tools. In: Current Trends in Web Engineering, pp. 320–324. Springer International Publishing, Aalborg, Denmark (2013)
61. Zbick, J., Jansen, M., Milrad, M.: Towards a web-based framework to support end-user programming of mobile learning activities. In: 2014 IEEE 14th International Conference on Advanced Learning Technologies (ICALT), pp. 204–208. IEEE, Athens, Greece (2014)

62. Zbick, J., Nake, I., Jansen, M., Milrad, M.: mLearn4web: a web-based framework to design and deploy cross-platform mobile applications. In: Proceedings of the 13th International Conference on Mobile and Ubiquitous Multimedia MUM'14, pp. 252–255. ACM, Melbourne, Australia (2014)
63. Zbick, J., Nake, I., Jansen, M., Milrad, M.: A web-based framework to design and deploy mobile learning activities: Evaluating its usability, learnability and acceptance. In: 2015 IEEE 15th International Conference on Advanced Learning Technologies (ICALT), pp. 88–92. IEEE, Hualien, Taiwan (2015)
64. Spikol, D.: CSCL Opportunities with digital fabrication through learning analytics (Poster). In: Proceedings of the 11th International Conference on Computer Supported Collaborative Learning (CSCL) (2015)
65. Gil de La Iglesia, D., Sollervall, H., Zbick, J., Delgado, Y.R., Sirvent Mazarico, C.: Combining web and mobile technologies to support sustainable activity design in education. In: Proceedings of the Orchestrated Collaborative Classroom Workshop 2015, pp. 1–4

Author Index

A
Akahori, Kanji, 1

C
Cárdenas, Leonor, 55

E
Ebner, Martin, 121, 139

G
Göktaş, özlem, 139

J
Jansen, Marc, 167, 193
Jung, Yong Ju, 101

K
Khalil, Hanan, 121
Kitazawa, Takeshi, 1

L
Land, Susan M., 101
Lucke, Ulrike, 23

M
Milrad, Marcelo, 167, 193
Moebert, Tobias, 23

P
Peña-Ayala, Alejandro, 55

S
Şad, Süleyman Nihat, 139
Sato, Koki, 1
Schön, Sandra, 121
Sotsenko, Alisa, 167
Spikol, Daniel, 193

V
Vogel, Bahtijar, 193

Z
Zbick, Janosch, 167, 193
Zender, Raphael, 23
Zimmerman, Heather Toomey, 101
Zuliani, Barbara, 121

© Springer International Publishing Switzerland 2016
A. Peña-Ayala (ed.), *Mobile, Ubiquitous, and Pervasive Learning*,
Advances in Intelligent Systems and Computing 406,
DOI 10.1007/978-3-319-26518-6